THE FAST AND THE FURIOUS: DRIVERS, SPEED CAMERAS AND CONTROL IN A RISK SOCIETY

Human Factors in Road and Rail Transport

Series Editors

Dr Lisa Dorn
*Director of the Driving Research Group, Department of Human Factors,
Cranfield University*

Dr Gerald Matthews
Professor of Psychology at the University of Cincinnati

Dr Ian Glendon
*Associate Professor of Psychology at Griffith University, Queensland,
and President of the Division of Traffic and Transportation Psychology
of the International Association of Applied Psychology*

Today's society must confront major land transport problems. The human and financial costs of vehicle accidents are increasing, with road traffic accidents predicted to become the third largest cause of death and injury across the world by 2020. Several social trends pose threats to safety, including increasing car ownership and traffic congestion, the increased complexity of the human-vehicle interface, the ageing of populations in the developed world, and a possible influx of young vehicle operators in the developing world.

Ashgate's 'Human Factors in Road and Rail Transport' series aims to make a timely contribution to these issues by focusing on the driver as a contributing causal agent in road and rail accidents. The series seeks to reflect the increasing demand for safe, efficient and economical land-based transport by reporting on the state-of-the-art science that may be applied to reduce vehicle collisions, improve the usability of vehicles and enhance the operator's wellbeing and satisfaction. It will do so by disseminating new theoretical and empirical research from specialists in the behavioural and allied disciplines, including traffic psychology, human factors and ergonomics.

The series captures topics such as driver behaviour, driver training, in-vehicle technology, driver health and driver assessment. Specially commissioned works from internationally recognised experts in the field will provide authoritative accounts of the leading approaches to this significant real-world problem.

The Fast and The Furious: Drivers, Speed Cameras and Control in a Risk Society

HELEN WELLS
Keele University, UK

ASHGATE

Published by
Ashgate Publishing Limited
Wey Court East
Union Road
Farnham
Surrey, GU9 7PT
England

Ashgate Publishing Company
Suite 420
101 Cherry Street
Burlington
VT 05401-4405
USA

www.ashgate.com

British Library Cataloguing in Publication Data
Wells, Helen.
   The fast and the furious : drivers, speed cameras and
   control in a risk society. -- (Human factors in road and rail transport)
   1. Automobile drivers--Attitudes. 2. Speed limits--Public
   opinion. 3. Speed traps--Public opinion. 4. Risk-taking
   (Psychology) 5. Liability for traffic accidents.
   I. Title II. Series
   363.1'256-dc22

ISBN: 978-1-4094-3089-6 (hbk)
ISBN: 978-1-4094-3090-2 (ebk)

Library of Congress Cataloging-in-Publication Data
Wells, Helen, 1977-
   The fast and the furious : drivers, speed cameras and control in a risk society / by Helen Wells.
      p. cm. -- (Human factors in road and rail transport)
   Includes bibliographical references and index.
   ISBN 978-1-4094-3089-6 (hbk) -- ISBN 978-1-4094-3090-2 (ebook) 1. Speed limits--Great Britain. 2. Automobile drivers--Great Britain--Psychology. 3. Photography in traffic engineering--Great Britain.
   4. Legal photography--Great Britain. 5. Traffic monitoring--Great Britain.
   6. Traffic safety--Great Britain. 7. Risk-taking (Psychology) I. Title.
   HE5663.A6W45 2011
   388.3'1440941--dc23

                                                                    2011030402

Printed and bound in Great Britain by the
MPG Books Group, UK.

Contents

Acknowledgements

I would like to thank Matthew Millings, Susanne Karstedt, Helen Beckett, Michael Fiddler and Bethan Loftus for their continued support and encouragement throughout this project and beyond. Also the ESRC for the funding that made the project possible and all the participants in the research who offered their views so enthusiastically.

Finally, to the little girl on her bike who, as I drove past her in my car, waved to warn me of the presence of a mobile speed camera ahead: thank you for reminding me why this was a topic that needed researching.

*To Matthew (who was there from the start)
and Daniel (who came as quickly as he could)*

Reviews for
The Fast and The Furious: Drivers, Speed Cameras and Control in a Risk Society

Helen Wells sets out on an important and timely quest to place roads policing through speed cameras in the context of a 'risk society'. Rightly, she avoids a debate about their effectiveness. Rather, she looks at changes in policing through greater use of technology and at the roles played by researchers, pressure groups and experts. As an expert cited, I found this a fascinating survey of a controversial topic.

Robert Gifford, Executive Director, Parliamentary Advisory Council for
Transport Safety, UK

We've waited a long time for this fresh perspective on a topic that touches us all in risk society. Using a multi-method, multi-site empirical study as her basis, Wells unpicks the many and contradictory strands of the speed camera debate, deliberately retaining a neutral stance and positioning the whole enterprise within a risk narrative. As such it delivers a powerful analysis of what was seen to "go wrong" through giving 'voice' to drivers, and serves up timely insights for the enforcing authorities. A real tour de force!

Claire Corbett, Brunel Law School, UK

A real thought provoker for anyone who has ever had an opinion about speed cameras! Through the voices of drivers, enforcers, persuaders, and decision-makers, this is an insightful look at the debate on arguably the most contentious of 'techno-fixes'. In explaining how, in many people's eyes, 'safety cameras' became 'speed traps', Wells reminds us that opinions cannot be changed by scientific evidence alone and that public acceptance is a prerequisite for any intervention.

Lindsey Simkins, Royal Society for the Prevention of Accidents, UK

Chapter 1

Introduction

This book does not seek to answer the question of whether or not speed cameras 'work', nor does it seek to contribute to the ongoing arguments for and against speed camera use on this basis. Indeed it deliberately side-steps such questions of effectiveness in favour of an approach that seeks to explain why this may be largely irrelevant given the individual and social context in which speed cameras are deployed and experienced. The work draws on various literatures to explain how they, in conjunction with the speed camera as the central iconic image in this road safety debate, provide the means, motive and opportunity for the debate that has been set in motion. The research focuses on the period of the life of the National Safety Camera Programme (NSCP) which operated between 2001 and 2007, but explores speed limit enforcement both before and after that period, up to and including the decision, in July 2010, to withdraw funding from local authorities with the knock-on effect of some cameras being switched off.[1] The book's aim is to suggest ways in which interested parties in road safety contexts and beyond can learn from the speed camera debate and can better understand the significance of the human factor, in this case the driver, in enforcement efforts that are largely dehumanised.

Since 1991, one approach to reducing the numbers of people killed or seriously injured in road crashes on UK roads has been to attempt to reduce the speed of traffic through the use of speed cameras. Between 1994 and 2006, the number of speed cameras[2] on UK roads grew from around 30 (Hooke et al. 1996) to over 5,000 (DfT 2009a), largely as a result of the existence of the NSCP which operated between 2001 and 2007. This programme allowed enforcing authorities to retain the fine revenue generated by speed cameras for reinvestment into further enforcement – a process known as 'hypothecation'. This was a controversial development and, despite the cessation of the programme, the UK's roads and drivers remain subject to unprecedented levels of enforcement of speed limits by speed camera. Between the launch of the NSCP and the end of 2008,[3] over 11.5 million offences of exceeding the prescribed speed limit on a public highway

---

1   And, in the case of Oxfordshire, back on again on 1 April 2011 (BBC News Online 2011a).

2   Including both fixed and mobile cameras. Figure represents most recently available data.

3   The last year for which data is currently available.

had been detected by speed cameras.[4] It is inevitable that every one of the UK's estimated 37 million drivers will have encountered speed camera surveillance at some point in their driving history, and that many more will continue to be rendered problematic by it in the future.

Enduring and, at times, heated discussions about the legitimacy and effectiveness of the use of speed cameras as a road safety intervention have accompanied their use and taken place in various realms during the last decade. From the Houses of Parliament to the editor's desk to the dinner table, everyone, it seems, has a view on the use of speed cameras and on the ways in which they should be financed, operated and deployed. As one recent review of the evidence about their effectiveness noted in its opening lines: '[if] there is one subject which divides drivers like no other it is speed cameras' (Glaister, in Allsop 2010, ii).

The word 'debate' will be used, throughout this book, to describe what has effectively been a public discussion of the legitimacy of the use of speed cameras as a means of securing the stated road safety objectives underpinning their use. At its most extreme, this debate has seen, in one corner, gangs of drivers firebombing, chain-sawing and otherwise vandalising speed cameras to put them out of use (see Motorists Against Detection 2006) with Safety Camera Partnerships having to consider installing CCTV to protect speed cameras (BBC News Online 2007a). In the opposite corner, it has seen groups of village residents blockading traffic to draw attention to their campaign for the installation of cameras, communities constructing their own replica cameras (see, for example, BBC News Online 2001, 2002a, 2003a, 2010a) and members of the public arming themselves with speed detection equipment (see, for example, BBC News Online 2005a, 2005b, 2006a, 2008).

Speed cameras have also been the focus of regular newspaper exposés (see, for example, *Daily Express* 2002, 2010a, 2010b; *Daily Mail* 2004, 2009; *Daily Telegraph* 2010a) and feature frequently in parliamentary questions. Speed cameras have also featured in advertisements for, among other things, cars, board games and mobile telephones, and have come to rival religion and politics as a subject of dinner-table conversation. This debate has endured despite the publication of numerous reports testifying to speed camera effectiveness (see, for example, Corbett 1995; Stradling 1997; Stradling and Campbell 2002; Buckingham 2003; Gains et al. 2004; Gains et al. 2005; Hirst, Mountain and Maher 2008;[5] Wilson et al. 2010; and see the studies referenced in Pilkington and Kinra 2005). Similarly, despite countless polls and surveys demonstrating that cameras are also popular, they remain a controversial intervention and continue to secure headlines and to animate politicians and the public alike.

---

4   It is not possible to ascertain how many *drivers* have been detected, given that this figure includes repeat offenders, but this figure is derived from data from Povey et al. 2010.

5   Mountain's research has been reported as being largely critical of speed camera effectiveness.

The research on which the book is based was carried out using a multi-method approach, drawing on media, expert, public and academic representations of and discussions about the use of speed cameras. It is based around interviews with the key 'expert' figures who have sculpted and re-sculpted the public image of speed cameras since their introduction to the UK's roads, as well as around focus groups with drivers themselves – the audience for this public relations exercise and those on the receiving end of the deployment of speed cameras. In addition to this, the work explores the media image of the speed camera, academic and other research works which contribute to the debate and the efforts of various pressure groups both for and against the use of the speed camera.

Attitudes to Speed Cameras, Speed Limits and Speed Limit Enforcement

A summary of survey evidence conducted over the core period of this research reveals some apparently contradictory findings:

- While 80 per cent (DfT 2003a, iv) of drivers have apparently expressed support for the use of speed cameras, 47 per cent 'would be either not very likely or not at all likely' to report someone vandalising one to the police (RAC Foundation report featured in Woodman 2003) and 54 per cent of drivers thought they made people drive more erratically (Datamonitor 2006, 13).
- Seventy-two per cent of drivers think that speeding in a 30 miles per hour (mph) zone is 'very serious' (Mori 2001) but 66 per cent of drivers consider speeds of 30 mph and over to be safe on such roads (National Statistics 2003, 7) and 31 per cent of all motorists agree that 'there are far too many speed cameras on Britain's roads' (Autonational Rescue 2009).
- Between 55 per cent (RAC Foundation 2005) and 99 per cent admit to speeding themselves (Corbett and Simon 1992, 74), 40 per cent of motorists have speeding convictions (Datamonitor 2006, 13), but 84 per cent consider themselves to be 'law-abiding' drivers (RAC Foundation 2005) and 60 per cent of men and 66 per cent of women say they conform to the speed limit (Stradling et al. 2008, 5).
- Drivers want to go faster (National Statistics 2003, 10) but also slower (Stradling et al. 2003, 108–9). Sixty-nine per cent want cameras to be visible (Mori 2001), although 'most' drivers believe that 'most' other drivers slow down for cameras and then speed up again (Stradling et al. 2003, 8).
- While '80 per cent of people questioned agree[d] with the statement that "cameras are meant to encourage drivers to keep to the limits not punish them"' (DfT 2003a, iv), 45 per cent thought that they were also 'an easy way of making money out of motorists' (ibid., 5–3) and it has been claimed

that support for the method has declined by 17 per cent between 1999 and 2009 (Institute of Advanced Motorists 2009).[6]

Clearly, not only are the topics of speed and speed cameras ones that have attracted considerable research attention, but the findings of these surveys generate some confusion in respect of the *real* story of public attitudes towards speed cameras. No doubt explanations for these differences could be suggested as a result of a detailed methodological critique of each finding, but such detail is seldom, if ever, presented alongside the reporting of such statistics, particularly when they appear in the media.

Although it is common for supporters of speed camera use to maintain that the majority of the public are in favour of speed cameras as a road safety intervention, and that objections are confined to a vocal minority (see, for example, Mountain 2008; Allsop 2010), previous research that has taken a qualitative survey approach has found that, while 2–3 per cent of drivers express outright hostility towards speed cameras,[7] only between 0 and 11 per cent of drivers expressed support for them[8] (Blincoe et al. 2006, 374). Findings such as these, and the contradictory findings of quantitative surveys, support the case for a genuinely open qualitative exploration of drivers' views towards speed cameras, which it seems cannot be adequately represented by a for/against dichotomy. Seemingly, while outright hatred of the method is limited, concerns of some kind do appear to be present, leading to support being qualified in many cases. For example, whilst it is common for surveys to ask members of the public if they 'support the use of speed cameras *as a road safety intervention*', agreement with the statement does not rule out drivers potentially doubting that cameras *are* used as a road safety intervention. Not all attitudes to enforcement by speed camera can be captured in survey research and, where qualitative contributions have been made alongside those sought out via surveys, research has tended to indicate a more complex picture and one that supports the research presented here (see, for example, Blincoe et al. 2006; Corbett et al. 2008; Soole, Lennon and Watson 2008). Issues of fairness, justice, distress, anger and frustration emerge as characteristic of at least some drivers' experiences of detection and prosecution by speed camera, and are seen to result in resentment towards the method. This is a worthy topic of consideration regardless of the proportions of people who can be shown to exhibit these feelings, given that any kind of opposition or reservations represent a challenge to regulation of this type.

It is not possible to account for the differences in the above research findings in any simple sense, and this work is not concerned with doing so. Instead,

---

6   For similar, apparently contradictory, statistics, see Musselwhite et al. (2010).

7   The research categorised responses according to Corbett's (2000) fourfold typology of drivers, finding that 0 per cent of defiers, 3 per cent of conformers, 2 per cent of the deterred and 2 per cent of manipulators expressed outright hostility to cameras.

8   Zero per cent of defiers, 6 per cent of conformers, 11 per cent of the deterred and 8 per cent of manipulators.

this research is concerned with two foci which, it is proposed, go some way to explaining both the persistence of the debate and the form it has taken. First, this work considers the production of expertise about speed limit enforcement (such as that reflected in the surveys above) and its use and abuse within what can be termed an expert 'marketplace' in which it is directly contradicted and challenged. Second, it considers the experience of the driver as the intended audience for the debate and the recipient of expert knowledge. The response of drivers to being presented with a variety of contradictory expertise about the topic of speed and speed cameras forms the second part of the research.

By taking the debate itself as its topic, this work seeks to demonstrate the importance of understanding the popular and public image of speed cameras, as well as the reasons for opposition to them where in evidence. The accuracy or otherwise of the risk calculations underpinning debates such as these is, it is argued, of less importance to the actual enforcement experience of both government, practitioners and drivers themselves than the methods through which enforcement is pursued. Hence, years of attempting to ask the question 'do they work?', and to publicise the results, have seemingly failed to remove the image of the speed camera from the pages of the national press, the attention of local and national news and internet-based message boards, not to mention the dinner table.

This focus is, I argue, particularly necessary given that many elements of the speed camera debate look set to be replicated in other contexts in the future. The use of 'risk' as the basis for enforcement, the use of automated surveillance technologies and the use of strict liability laws in combination with fixed penalty, out-of-court disposals are just a few of the significant components of this debate that the (motoring and non-motoring) public can expect to encounter in future years whether they be, for example, littering, allowing their dog to foul a public place, smoking in a proscribed location or parking incorrectly or beyond the limit of their ticket.

The research presented here should not be used as evidence in support of the argument against speed cameras but rather as an exploration of the challenges under which speed cameras are being operated. Whilst the findings reported here from drivers are, admittedly, largely negative, this is not intended to be taken as reflecting a random sample of those encountered. Rather, these findings are presented here as offering insight into, and potentially explanation for, why the deployment of this technology has been so undeniably controversial. While not claiming that these voices represent every motorist, or necessarily even the majority, the work seeks to unpick and analyse negative views about the use of speed cameras so that the use of technologies of control such as these in the future can be informed by this enforcement experience. Criticisms of the use of speed cameras are undeniably present within the debate around their use and, it is suggested, it does not make any sense to challenge these criticisms with the use of casualty statistics or conviction data. Instead, the social context in which these statistics are produced and witnessed is a necessary topic for understanding challenges and resistance to enforcement technologies. My exploration of this context does not indicate agreement with or support for these dissenting views, and this book should not be taken as a contribution to the argument against the use of speed cameras. It should, however, be taken as an invitation to engage with the arguments put forward by those who experience speed camera surveillance as drivers, and an endorsement of the benefits to be derived from trying to understand, not simply trying to suppress, oppositional voices. The benefits of doing so for this and other future road safety and wider enforcement campaigns are, I suggest, considerable. The final chapter considers ways in which this analysis can be used constructively with future risk reduction in mind.

Theoretical Framework

At a basic level, speed cameras can be seen as the end product of a process whereby a behaviour has been identified as 'risky' and therefore as increasing the chances of negative outcomes occurring. Such a negative outcome is, in this case, road death and injury. As has been suggested, this debate appears to be, at least on one level, concerned with the deceptively simple question of whether or not speed cameras 'work' in terms of reducing the incidence of such harms. However, such a question of effectiveness contains within it a multitude of other questions. These relate to:

1. the accuracy with which the causal chains linking speed and the occurrence of harm have been observed
2. the probability of such causal chains resulting in harm, and
3. the appropriateness of the speed camera as an intervention in this process.

As such the deployment of speed cameras is motivated by a concern to reduce 'risk', but, this work argues, risk is also of significant importance in creating the *regulatory context*, dictating the *form* of intervention and (crucially) generating the *reaction* to it. The debate is therefore viewed as a product of a society in which multiple discourses based around a concern with 'risk' are used by both individuals and organisations to structure their understanding of the issues of speed and speed enforcement. Such discourses are, it is proposed, used in generating both the policy and the reaction to it, and provide an explanation for both the existence and the intransigence of this particular debate.

This debate, it is proposed, takes place not simply in a 'risk society', but in 'a society of risk discourses' (Hunt 2003, 183) where risk is never 'technically neutral' but always a 'moralised form of government' (O'Malley 2004, 326) and in which the cultural construction of risk is central to understanding how it is used and understood by the various individuals and organisations who compete over it. As such, the primary emergent meanings of risk in evidence here are drawn from what can be described as the actuarially informed conception of risk (often drawn upon by those in favour of the use of speed cameras) and a more culturally informed notion, the latter offering the context in which the 'evidence-based' vision of the benefits of speed cameras is rendered problematic by the cultural context in which their deployment takes place.

Speeding and speed enforcement are not *big* risks in the sense of Beck's examples of nuclear annihilation or environmental devastation (1992), but are instead viewed as 'everyday risks' – directly encountered and experienced on a daily basis (Hunt 2003, 167). At one level, this debate has been conducted in terms of apparently measurable causes and effects based around the effectiveness of speed cameras at reducing the negative side effects of mass individual mobility. This book, however, attempts to explore an entirely different set of 'side effects' (ibid., 171) based around the 'social consequences' of scientific innovations (ibid., 184). As such, it is, to use Haggerty's phrase, an empirical study of control technologies in use in risk society (2004a, 493). Such studies are important given that, it is argued:

> [T]echnology should not be seen as consisting of a physical, material dimension only; rather, technology operates in a social context and its meaning is perceived differently by people in different social and organizational positions. While technological changes have the capacity to transform social organizational life, technology is itself shaped by social and organizational conditions. (Chan 2003, 668–9)

Within the risk literature, the 'traffic' context within which this work is situated often features as an example of a risk which fails to excite much interest. Traffic risks are often viewed as 'natural', 'inevitable' and 'legitimated' (O'Connell 1998, 114), assimilated into our daily routines in a way that deprives them of the fear associated with many risk issues: 'If you are going to jaywalk, look both ways before you step into the street, and if the street is busy, be nippy. Many risks are ignored, after routine safety measures are taken. Driving a car, for instance, is dangerous, but few of us worry about it' (Hacking 2003, 22). As such, the risks associated with road use generally feature as the exception to the rule, as the example that fails to get people (authors or subjects) very animated. Traffic is, instead, used to illustrate the manageability of risks: '[T]he threat of crime has become a routine part of modern consciousness, an everyday risk to be assessed and managed in much the same way that we deal with road traffic – another modern danger which has been routinized and "normalized" over time' (Garland 1996, 446).

When acknowledged as a risk, road risk is something to which 'we' are exposed by external others, and which 'we' must learn to avoid: 'The risks we run depend on the actions of others and the risks they take. Steering clear of dangerous drivers is the simplest example' (Garland 2003, 55). The prioritisation of speeding as an enforcement target, however, has increased the salience of this kind of traditionally desensitised risk issue to an unprecedented degree. Corbett has noted that criminology has failed to engage with road policing/offending generally, despite the fact that road policing constitutes 'the most common means of police-initiated contact with citizens' (2008, 132). This increased profile has also allowed the question of road risk to be viewed from an alternative perspective, whereby road risk becomes something 'we' cause rather than something 'we' are only at risk of. The debate is therefore considered to be about much more than the effectiveness of a deployed technology at reducing risk, although such questions frequently surface within it. This work centres its explanations for the endurance and form of the debate in 'risk' literature, broadly understood, and suggests that the speed camera, the iconic symbol of this debate, can be used as a lens through which questions of truth, of identity, of regulation and of justice, in a society increasingly prone to conceptualise problems in 'risk' terms, can be explored.

Methodological Approach

Much research has been concerned with answering various interpretations of the question of whether or not speed cameras 'work' (see, for example, Corbett 1995; Stradling 1997; Stradling and Campbell 2002; Buckingham 2003; Gains et al. 2004; Gains et al. 2005; Hirst, Mountain and Maher 2008,[9] and see the studies referenced in Pilkington and Kinra, 2005). Despite this effort, the question continues to surface within the debate and seems peculiarly resistant to being answered categorically. Rather than trying to address this question, this book explores the wider societal issues that explain why this question is unlikely ever to be answered in a way that will resolve the debate or put an end to discussions about speed camera use. It will also demonstrate why, to a large extent, such a question obscures the wider issues that lead to the perpetuation of the debate.

The research therefore concerned itself with exploring the context in which this question continually resurfaces, and operated with a number of research questions. These included asking, at a theoretical level, to what extent this debate was informed and influenced by its location within a society increasingly concerned with risk, including asking how the acceptability of control forms such as the speed camera are affected by the 'risk' context in which they are experienced. The study also asks, however, what this debate can tell us about wider issues of identity, justice, regulation and the production of knowledge/truth in societies concerned with risk in many senses of the word. In this final sense, the work transcends the case study around which it is based and allows for suggestions to be made about other contexts (both within and beyond that of the road/driver) where restriction, control and (potentially) punishment are legitimized by a concern with the minimisation of risks.

The debate about the use of speed cameras is therefore the subject of this analysis, with the speed camera providing the lens through which this particular risk

---

9   Mountain's research has been reported as being largely critical of speed camera effectiveness.

debate can be viewed. The final chapter of the book, however, makes observations and recommendations transferable to other contexts with an apparently similar profile which are of interest to both those charged with reducing risk and those who become the focus of attention because of this.

To fulfil these various aims, a multiple-site, multiple-method approach has been used in this research in order to access the various arenas and forums in which a public discussion about speed cameras takes place. These arenas are seen to feature both verbal and printed contributions to the debate, and are located in both the real and the virtual world. The participants whose input is sought in the production of an assessment and exploration of the debate are, additionally, seen to come from official, quasi-official and independent backgrounds and to use a variety of outlets for communicating their views. As a result, this chapter describes the interview, focus group, observation, documentary analysis, media review and internet-based methodologies used in order to access each aspect of the debate considered significant for the purposes of this research and in order to address the research questions set out below.

Research Questions and Approach

The purpose of this work is to study the debate around the use of speed cameras as speed limit enforcement technologies under the auspices of the NSCP, rather than contribute to that same debate. As such, a number of research questions have been formulated:

1. Why has a debate occurred around the enforcement of speed limits by speed camera?
2. What forms has the debate taken and who has engaged in it?
3. To what extent is the debate informed and influenced by its location within a society increasingly concerned with matters of 'risk'?
4. What strategies and tactics of control are used in this debate, and what is their contribution to the debate?
5. How is the acceptability of such control affected by the risk context in which it is experienced?
6. What can this debate tell us about wider issues of identity, justice, regulation and the production of knowledge/truth in societies concerned with 'risk'?

Given this focus, the methodology used here is substantially different from that which underpins the debate itself, based as it is around quantitative measures of camera operation, aggregate data of road crashes and their causes, and surveys and opinion polls of the popularity of speed cameras. Although such quantified data is a key element *of* the debate about speed enforcement, it is not appropriate for research located outside of that same debate, and which therefore takes that debate as its subject. Although the debate is a forum for the exchange of largely quantified, aggregate data, this exchange is itself conducted with words, via conversations, publications, press releases and other qualitative forms of exchange. Such qualitative input is designed ultimately to convince the various audiences of the legitimacy of the scheme or to generate opposition to it.

Rather than taking polls of the type summarised previously, or any other research evidence, at face value, this research attempts to understand the fact of their existence, their inability to resolve the debate and their wider meaning for both those who have produced them and those for whom they have been produced. Ultimately, it uses the particular theoretical framework of 'risk' to explain not only why a debate of this nature can flourish and defy attempts to resolve it, but why attempts are continually made to do just that. The research therefore considers both the experiences of policy *making* and policy *taking* in relation to speed limit enforcement by speed camera and in a society in which both processes are influenced by a concern with risk.

This methodological approach recommends two levels at which the debate occurs: the 'expert' level and the 'public' level. In the first instance, the production of, and participation in, the debate at organisational and official levels is a focus for research attention. In this sense, it is the interplay between the different key voices that is of interest, combined with their many and varied ways of negotiating and maintaining a role in the debate, and their beliefs and assumptions about their audience. The way in which each aspect of their expertise and status is affected by the increasingly risk-dominated context in which they operate is a further area of research. This focus recommends a number of subjects, areas of interest and types of data drawn from a variety of debating arenas, and necessitating a variety of methodological approaches including:

- a literature review and documentary analysis of academic texts and policy documents
- a review of the public internet profiles of the expert voices
- interviews with the key expert voices within the debate
- a review and analysis of national and local press coverage.

At a second level, the intended audience for this debate is of significant interest. In this sense, the experience of being faced with a debate among experts who disagree (one that nonetheless ultimately shapes the reality to which the public are subjected) is of interest. This experience, and the reflexive involvement of the public in perpetuating the debate themselves, forms the second tier of debate, and allows for behaviour often represented as simply 'taking sides' to be explored for all its contingencies and conditions. This, again, necessitates a variety of methodological approaches including:

- pilot research based upon the collection of postings from web-based message boards

- the observation of an alternative disposal for speeding drivers – the Speed Awareness Course
- focus groups with drivers with different 'driving biographies'.

Although the research cannot, therefore, claim to be representative of the general driving population, the methodology used provided participants with the opportunity to provide, in detail, their feelings about and attitudes towards the issues, without the need to restrict themselves as would be the case with a questionnaire-type approach using closed-format questions. This type of qualitative research is deemed particularly suitable for exploring the views of individuals who have experienced a system notable for the absence of opportunities to 'voice' (being a process initiated by a machine and prosecuted by post), allowing complex issues to emerge which would not necessarily be accessed via the prescriptive form of a questionnaire, where topics are predetermined (in large part) by the author of the instrument.

Documentary Analysis: 'Front-Stage' Expertise

The ways in which organisations, interest groups and agencies choose to represent themselves via text (on paper or electronically) are a vital constituent of the way they choose to portray themselves as expert witnesses to the issue of speed. Documents produced as material for supporting a particular debating position are thus efforts at 'self-presentation' which are strategically constructed (Atkinson and Coffey 2002, 45).

Strategies, press releases, statements and policy documents are 'among the methods whereby organisations publicise themselves, compete with others in the same market, justify themselves to clients, shareholders, boards of governors or employees' (ibid., 46). In this sense, they can be seen as competitors in the market for supplying 'truth' to the public and to other experts and agents in the field. They are at the same time engaged in 'persuasive work', competing to provide the most convincing interpretation of what is going on, and they are 'justifying' texts (Watson 2002, 80). They appeal to other experts, to the media and to members of the public as an electorate, as members of interest groups, as offenders, as victims and as taxpayers.

Such documents are the front-stage evidence of the debate (Goffman 1959/1990), the research, the commentary and the policy designed for public consumption. According to many of these documents, however, there is no debate. Each is published as an unchallenged text, as the considered, strategically constructed and deployed 'case' of each side. As such, these documents are a vital starting point through which to view the debaters' public face. Once fed into the debate, however, they become a focus for the attention of debate opponents: dissected, critiqued and challenged in order to undermine opposing stances.

This analysis involved the examination of strategies, policy statements, consultations and research reports issued in support of the current policy of using speed cameras to enforce speed limits. Such documents are published by the traditional expert voices of central government and are an opportunity for them to present a definitive case in the debate. It is through these documents that they argue for the mandate to intervene in risk issues in certain ways. This analysis therefore considers the language and presentation, as well as the evidence base and academic content, of relevant policy documents. Particular emphasis is placed on those produced since 1991 and referenced in the policy documents that set out the case for a national safety camera scheme.

Press releases are issued frequently by agencies and organisations located on both the supportive and oppositional 'sides' of the debate, in particular by the Department for Transport, the Home Office and by motorists' organisations. Press releases offer the opportunity to present, unchallenged, one side's views or research, and are opportunities to present smaller, simpler and more digestible versions of larger strategies, manifestos and position documents. As such, they are a valuable method through the use of which the debating experts can hope to make their expert product bridge the gap between the expert debate and the lay audience for whom it is designed. The choice of tone, language and approach, the presentation of complex research and the construction of the 'sound bite' are all significant tactics for communicating with the audience, whether that is envisaged to be the general public, other experts or even the media itself.

Academic Texts

Academic and other research texts focusing on the effectiveness of speed cameras are of interest in this debate context more for the very fact of their existence and their subsequent co-opting into the debate than for their actual contribution to it. Their conclusions and recommendations are therefore not as significant for this research as the way in which they are subsequently used within the debate to support a different stance or to evidence a particular approach.

As such, this study has considered what various sociologists, criminologists, transport analysts, road safety specialists, engineers, historians and psychologists have contributed to the ongoing debate about speed enforcement since its earliest manifestations at the start of the twentieth century. Again, particular attention has been paid to the ways in which this product has been increasingly commandeered to legitimate and justify policy, as well as to construct challenges to it. Particular emphasis is again placed on those produced since 1991 and/or referenced in the policy documents that set out the case for a national safety camera scheme.

The Media Realm: Review of Local and National Press Coverage

The various media operate both as a key source of information for the debate and as one of its key battlegrounds. They are a source both for other experts and for the public, and an outlet via which the prepared statements and publications of the expert agencies and agents can be publicised. Beyond this, the various media types are also active voices in their own right, selecting what they report and how they choose to present it, with their specific agendas and political sympathies determining how hospitable each source is to each expert's product.

As such, the media dictates the form of output of the other experts to a large degree, given its importance as an access point to the debate for the general public. The degree to which it shapes, as opposed to reflects, public opinion is, of course, in itself debatable. The media review conducted as part of this research consisted of a review of a sample of newspaper articles and online BBC news stories over the period since the NSCP was launched and beyond. This was not a complete review of every story related to speed enforcement produced during the research period, but represents a sample of those contained within the newspapers and sites mentioned. It is intended to generate an impression of the positions of the various media, rather than be taken as a comprehensive survey of all their output. National and local press coverage was analysed, including both broadsheet and tabloid representations and national and local BBC reporting.

A total of 12 newspapers were included in the sample, with articles selected that related specifically to the NSCP as in operation at the time. These approaches produced a total of 176 articles that were specifically related to or mentioned the NSCP and/or speed cameras as an enforcement technology. This sample was subsequently analysed in terms of, for example, the supportive or oppositional stance adopted, the arguments used and the use of expert sources. References made within articles to various expert voices were subsequently used in the generation of a list of potential interviewees for this aspect of the research (see below).

Interviews: 'Back-Stage' Expertise

The debate around the use of speed cameras to enforce compliance with speed limits is seen to result in large part from the interactions and efforts of a number of individual expert voices, representing the key organisations of the debate. Many of these individual agent representatives have become figureheads synonymous with the debate, more accessible than the official bodies they represent and, ultimately, potentially held responsible for the gains and losses associated with the policy. Accessing their experience of participation in the debate necessitates a strategy whereby these key individual actors are approached *as debaters* and not asked to engage further in the debate, or to recount the front-stage performances and 'impression management' (Goffman 1959/1990, 233) designed for public consumption. I do not mean to suggest that, by conducting interviews, I was able to access the unfiltered, unmanufactured back-stage performances of these participants. However, such interviews at least allowed access to a lesser level of performance than that required in public and provided some insight into the way in which such front-stage performances were 'prepared' (ibid., 231).

The final list of those interviewed was generated through a two-stage selection process comprising the foregoing media review and documentary analysis process and 'referrals' by other experts during interviews and communications with them. A 'referral' in this sense means that another interviewee specifically mentioned them as another voice within the debate, and was considered to include both positive (of the type 'you really should speak to', or 'we work alongside X') and negative (of the type 'they are really a problem for us', or 'the X talk a lot of rubbish') referrals. The mentioning, unprompted, of other voices, even to undermine or be derogatory about them, was considered to indicate a significant role for that voice, given the nature of the debate.

The second stage of the selection process considered the use of expert voices in the media, with each article analysed for its use of expert sources. There was a strong degree of correlation between the list of experts generated by the referral process and the list produced from the media review, with the first nine most-referenced experts from the media review also being recommended through the referral process used in interviews. Ninety-five per cent of news stories analysed contained a reference to at least one of these agents, 36 per cent referring to two, 17 per cent using three and a further 15 per cent using between four and seven. This resulted in a total of 387 uses of an expert reference in the 176 articles used for the review.

This list therefore represented the key 'voices' heard within the debate, arguing in favour of and against the use of speed cameras and in relation to speed management policies more generally. Interview participants were either approached in person (where that individual was considered to be a significant figurehead), or the agencies and organisations identified as significant were approached and volunteered a suitable interviewee.

A semi-structured interview style was used, somewhat akin to Holstein and Gubrium's 'semi-formal guided conversations' (2002, 113), with interview schedules tailored to the particular interviewee. Care was necessarily taken to adopt a style of interviewing that did not imply a stance in relation to speed enforcement, given the polarised nature of the debate and the potential for an opposing stance to alienate the interviewee. This stance was considered necessary to avoid receiving only the same front-stage performances that were used in conducting the debate on a more public level and for fear of eliciting the practised and rehearsed responses designed for persuasion in these more public spheres. Without this approach, it is unlikely that many interviewees would have admitted that a debate was actually taking place, and would not have been prepared to engage in conversations about the nature of their role within the debate. In some cases, it was only at the second interview that these kinds of conversations were possible, once a relationship of at least partial trust had been established.[10]

Some authors have suggested that there are advantages to appearing as a non-expert and allowing the interviewee to enlighten you, giving them a feeling of power (Miller and Glasner 2002, 105). This was not an appropriate strategy in interviews with the experts of this debate, as being a non-expert in the subject of the policy renders oneself a part of the audience and more likely to be subjected to the rehearsed debate. A non-expert is a potential recruit to a particular 'side', and adopting this position would have rendered me particularly vulnerable to attempts at persuasion, rather than allowed me access to the back-stage world of the expert. Demonstrating knowledge of the technicalities of the subject, for instance by using acronyms myself or displaying an understanding when interviewees used them, was the approach used instead. One senior government interviewee confirmed that this approach had been appropriate in terms of accessing the desired information, admitting that he assumed that I 'would be some person off the street' wanting to 'talk about *Daily Mail* headlines' and 'revisit the usual debates'. He visibly relaxed and 'opened up', becoming more involved and interested once I had used phrases such as 'legacy sites'[11] and 'the criteria'[12] or mentioned other people with whom I had spoken. These phrases seemed to establish that I was neither wasting his time nor wishing to engage in an argument, and he became less defensive and dropped his front-stage performance, even acknowledging that he had done so. I believe my approach helped me establish myself as a sympathetic person with whom he, and others like him, could share their concerns and frustrations about the debate, not someone for whom the usual 'familiar narrative constructs' (ibid., 101) would be sufficient.

The issue of anonymisation of the interview participants has been a complex one. Many respondents expressed a desire to be identified by name in my research, and I have effectively gone against their wishes in choosing not to do this. I have adopted this approach partly because the consent for (and, in some cases, desire for) the content of the interview to be attributed to them personally was obtained at the *start* of the interview. As the above example shows, the subsequent discussion sometimes took a slightly different course to that which the interviewee had originally envisaged participating in, and their consent for the attribution of

---

10   The researcher's completion of a secondment at the Parliamentary Office of Science and Technology provided the opportunity for a first round of interviews resulting in a publication (Wells 2004). Following this publication, a second set of interviews was arranged for the specific purposes of this research and it is considered that the publication's avoidance of controversy and of adopting a partial position helped with the arrangement and substance of the second round of interviews.

11   The phrase used to describe speed camera sites that pre-date the use of strict criteria for their placement.

12   The rules and guidance relating to speed camera placement, known within the debate as 'the criteria'.

the subsequent discussion *directly to them* was not reaffirmed at the end of the interview (although the agreement that I could use the interview in my research more generally was).

In a second sense, this wish to be credited by name was in some cases motivated by a desire to be promoted further within the debate, and to put forward their case in an independent, academic piece of work. For debating experts, this is (as will be discussed in Chapters 3 and 4) an opportunity not to be passed up, and my decision to anonymise my respondents is also partly motivated by a desire to distance my work from this self-promotion. This presents a difficult dilemma when an expert is also represented in publicly accessible forms, for instance in newspaper articles. Where this is the case, I have maintained the anonymity of the interviewee by avoiding explicitly juxtaposing the two in a way that could lead to the interviewee being traced.[13]

As a final practical note in relation to interviews, it should be noted that the location of each interview was dictated by the subject. This meant that, in many cases, the individual's office was used, but interviews also took place in the foyer of a London hotel (Chief Constable 2) and a motorway service station (anti-camera website author). Wherever possible, recording equipment was used with the consent of the interviewee and the interviews were later transcribed by the author.

The Virtual Realm: Internet-Based Research

In addition to, and alongside, the expert and media realms, the internet functions as an active arena in which the debate is conducted. Cornick notes that '[c]yberspace makes researchers of us all' (1997), and in a context such as this, where there is public debate about the issue of speed enforcement by speed camera, the internet is a readily available source of information for the debate's audience. It is therefore an important focus of this research. Given this function, it also forms an important outlet for the profiles constructed by the debate's experts, designed for consumption by a wide variety of audiences. The role and contribution of this virtual realm to the debate has been researched in three forms. First, the web-based output of the key experts was explored. As well as an outlet for promotional material in the form of press releases, research evidence and policies, for example, the internet allows for the various agencies and agents to present themselves in an unmediated form and with their profile constructed and controlled as they wish it to be received. Websites for the key players are repositories for all that that expert wishes to put forward into the debate. They serve to make this profile and all it

---

13   This has not proved possible with material accredited to 'Captain Gatso', given that MAD was effectively the only organised group of vandals operating at this time and any attempt to anonymise him would therefore be something of a token gesture.

contains as freely available as it can be made, but without the inevitable filtration and mediation that the news media brings with it.

Every expert agency or organisation that featured in the documentary analysis and interview aspects of this research had some kind of web presence where it could present its evidence, offer links to other sites, provide information or guidance and undermine other positions within the debate. Analysis of these alternative methods of self-presentation and self-promotion thus followed a similar format to the documentary analysis of more traditional textual sources (above).

The second area of virtual exploration was message boards. These boards (and their near neighbours, 'discussion forums' and 'newsgroups') offer members of the debate's audience the opportunity to engage in discussions about the policy which will be publicly available and which will be read by other interested parties, potentially including the 'experts' themselves. The boards used in this research were set up by a mediator (on behalf of the host site, in this case the BBC) in response to a particular news item about speed and speed limit enforcement, and were both national and local in their coverage. BBC forums could be accessed by anyone reading the relevant news story, who would be invited to 'Have Your Say' and respond with their thoughts about that story, many of which would be subsequently published on the website and invite further comments and responses. This aspect of the research was conducted over an 18-month period and the sites that feature in this research are those that were operational during this time. Selected sites and discussions were those that explicitly invited responses to issues related to speed and speed enforcement, and a mixture of national and local sites was selected. Local sites with different enforcement profiles were selected, for example the Welsh BBC site, representing an area where enforcement was particularly energetically pursued and which had achieved a certain notoriety within the debate. Message boards generally ran for a limited period and were then closed to further contributions. At this point, their content was saved electronically as well as in hard copy, and analysis of the messages was conducted. This analysis involved searching for specific terms and words as well as themes such as comparisons with 'real' crime and suggestions of a financial motive for enforcement. Because of restrictions on the use of BBC content in other publications, this research has been used only to suggest themes which could be pursued in more depth in focus groups and interviews, and direct quotations of this material will not feature here, although analysis has supported those findings obtained during focus group research and the reader is encouraged to visit sites of this nature.[14]

Third, the internet offers a variety of other speed- and speeding-related features of potential interest to Cornick's 'researchers' (Cornick 1997). Advertisements for products (both legal and illegal) that affect the operation of cameras, hide the identity of cars or detect the presence of cameras were numerous and

---

14   For example: news.bbc.co.uk/go/pr/fr/-/1/hi/talking_point/3306405.stm; news. bbc.co.uk/go/pr/fr/-/1/hi/talking_point/3390665.stm; news.bbc.co.uk/go/pr/fr/-/1/hi/ talking_point/3423425.stm; news.bbc.co.uk/go/pr/fr/-/1/hi/talking_point/3731585.stm

often contained product reviews in which their effectiveness was discussed. Additionally, many sites offer legal advice for detected speeders (such as their rights and obligations) and methods of avoiding effective prosecution (such as successful excuses, pro-forma letters for sending to the authorities, contact details for drivers willing to accept penalty points on behalf of other drivers). These contribute to the perception of speeding that the audience may have, and to their views about its impact on them, as well as providing information that can influence future behaviour. Sites and advertisements of this type are often the first to be listed when any speeding-related search is entered into a search engine. For example, the phrase 'speed camera' entered into the Google search engine produces links for 'motoring lawyers' (who will help you 'fight for your driving licence'), offers to 'beat speeding tickets', 'cancel your speeding fine' and 'get speed tickets cancelled', as well as numerous products that interfere with the effective functioning of speed cameras.[15]

As such, the debate is conducted at least partly in this virtual realm and it is considered to be a justified arena for research, not simply the 'easy option' in terms of collecting data (Illingworth 2001, para. 9.5). In addition to the uses and forms given above, the internet was also frequently volunteered as an important source of information in 'offline' conversations during the research, in a similar way to that in which media stories were sometimes drawn upon and used to illustrate and support opinions. Combined with the research methodologies detailed above, internet-based research offers an alternative, additional way into the expert-level debate and its forms and forums. Combined with the research detailed below, the internet offers an alternative way of accessing the public-level debate among lay members of the audience. This research could not have claimed to be anywhere near a full account or exploration of the debate if the internet had not been used as source of information or treated as a significant research site in its own right.

Speed Awareness Course Observation

Over 20 sessions of a Speed Awareness Course were attended between the launch of the scheme in 2003 and 2007, and then again in 2010. Observation was conducted as a fully overt non-participant observer and involved observing the scheme in action and speaking to participants and course leaders. Each course was attended by up to 20 drivers who had been detected speeding by a speed camera, but who had been offered the opportunity to attend an education course at their own expense in exchange for avoiding the addition of three penalty points to their licences. The courses were considered important, given that they offered a forum for direct communication between enforcers and the public. In bringing

---

15   This particular search was carried out on 19 September 2006, but the results remained similar in tone and content throughout the research period.

together representatives of the enforcing authority and drivers, they were also effectively bringing together the two separate debating spheres and offered an opportunity for public and expert voices to interact and challenge each other. The purpose of these observations was to identify how a development of this nature was received by those drivers who experienced it, including how they engaged with the subject matter and the format and, more generally, how they spoke about speed enforcement in the presence of other drivers and representatives of the authority pursuing the enforcement. Additionally, these observations provided further examples of the range of identities volunteered by drivers when invited to introduce themselves at the beginning of each course. This provided further reinforcement for the categorisation of focus group contributors with whom I could explore issues in more depth.

Each course was convened by a course leader and lasted a full day. The morning of each course comprised a number of educational and discussion exercises, whilst (for the first observation period) the afternoon was occupied by an 'on-the-road' exercise in which drivers were able to apply the knowledge they had acquired in the morning session.[16] At each observed session, therefore, the attendees were different, but the same course leaders were observed on numerous occasions. On each occasion I was introduced as a researcher and sat at the back of the room overtly taking notes on proceedings. A number of attendees asked questions about my research during the coffee and lunch breaks, which were answered fully. Conversations engaged in during break times were also noted, with the consent of those involved. Data was collected in note form during and immediately after each session, with occasional quotations transcribed in full.

Focus Groups

Given that speed enforcement is something of a 'dinner table' discussion topic, the focus group method was considered particularly suitable for accessing the form and nature of debate among ordinary members of the public. Focus groups enabled the in-depth comments and criticisms of individuals to be accessed and also for debates to be played out in an observed situation in which I was able to pose questions and to exert some influence over the questions discussed. Given that the focus group element of the research followed on chronologically from the other two methodologies designed to access the public level of debate (message boards and observation), focus groups were used as a venue for exploring some of the issues that surfaced in the more constrained environments of the message board and the observations, where my freedom to pose follow-up questions and to influence the types of topic discussed was more limited. As such, focus groups are

---

16   This element was dropped from the later courses as it was felt to be a deterrent to some drivers accepting the offer of the course; this also meant that the cost of the course could be reduced.

used to 'clarify, extend, qualify or contest findings on the same topic produced by other methods' (Bloor et al. 2001, 90).

Although certain characteristics are traditionally used to categorise and structure focus groups, traditional demographic categories were not used in this research. Discourses of truth, identity and justice are the focus of this research, not those specific to traditional cleavages of race, age, class or gender, and such identities were, in any case, notably absent when contributors were free to suggest their own identifying characteristics. The themes around which my focus groups were convened instead drew inspiration from the independently volunteered aspects of identity suggested as relevant by the previous two methodological approaches. The identities of professional driver, experienced driver, new driver and convicted driver presented as the most significant and were adopted as the categorisations for the focus groups. It was considered that the aspects of identity independently generated by participants and contributors offered more interesting and potentially fruitful lines of enquiry than the more traditional categories based on age and gender, for example.

Participants were selected via a purposive sampling strategy, guided by the particular areas of interest suggested in the prior research. It is one of the advantages of research into a relatively common offence that convicted speeders could be recruited relatively easily via snowball sampling methods and through a process of self-referral from among drivers known to me and their acquaintances. New drivers were recruited via a student association in which I was involved, while experienced drivers (defined as those with ten or more years' experience) were drawn from social circles and transport hobby groups in which I am involved. In the case of professional drivers, recruitment involved both contacting employers of professional drivers (haulage and removal companies, for example) and approaching professional drivers' unions and asking for willing participants. In this second case, I was able to make use of existing meeting arrangements, conducting a focus group once the union business had been completed.

Naturally, these groups are not based on exclusive categories and participants emerged as, for example, both experienced *and* convicted, professional *and* convicted. Where this was apparent before the groups took place, potential participants were invited to attend whichever groups they qualified for and to select the one they felt most accurately represented their own driving identity. Identities are anonymised within the research, but with some indication of age and gender to distinguish between participants. The theme of the focus group is also indicated – for example, '(Male, late 50s, professional driver focus group)'. Where a quoted passage includes my own input, this is indicated via the initials 'HW'. Selected quotations are presented here, and whilst not all focus groups were exclusively negative in tone, where support for cameras was expressed this was in all cases qualified, with some reservations voiced. Again, this is seen to further justify the use of the method as opposed to the more common survey approach where simple 'yes/no', 'support/oppose' dichotomies are often presented but fail to capture the subtleties that qualify and contextualise attitudes to the policy. The author's agreement with these views is not to be inferred from their selection and presentation here; rather, the intention is to demonstrate the nature and form of the regulatory challenge that they pose.

Focus groups ranged in size from three to eight people, and semi-structured questionnaire formats were used to initiate and steer discussions. Questions explored issues such as experiences of enforcement, preferred methods of enforcement, sourcing of information on speed and speed enforcement, and understanding of and participation in the debate. Where a response to recruitment under a particular category was sufficient, multiple groups were convened.

The Structure of the Book

This work considers the debate that has taken place around speed limit enforcement at both expert and lay levels. Chapter 2 explores the issue of speed limit enforcement throughout its history, as well as unpicking the regulatory framework that makes the current practice of speed camera use legally and technically possible. Some of the research data that has legitimated the use and spread of speed cameras is also considered. The focus then turns, in Chapters 3 and 4, to the concept of *expertise*, before Chapters 5 and 6 consider the importance of *experience*.

Chapter 3 explores how the debate around the use of speed cameras is a debate shaped by the risk context in which it operates, taking changes to the meaning and form of expertise as its focus. The significance of the 'demonopolization' of expertise, as described by Beck (1992, 29), is considered in relation to the causal chains promoted as underpinning the use of speed cameras. Expertise is seen to have become not only demonopolised but dispersed across a wider range of sources and to have incorporated a variety of newly enfranchised expert voices. The chapter therefore considers the key expert voices which are present within the speed limit enforcement debate and the two 'sides' that compete both to define the 'truth' of the problem of road death and injury and to obtain the mandate to intervene to reduce it. The chapter also explores each key debater's 'case' for being permitted to dictate the road safety reality in relation to speed.

Chapter 4 then considers the consequences of this demonopolisation both for those engaged in competing for definitive expert status in risk society and for the audience faced with a situation in which experts are unable to agree. The chapter proposes that a range of marketing strategies can be discerned within the expert debate, with competing experts using a variety of methods to promote their own interpretation of the causal chains at work over those of other experts. As a further aspect of the changes to the concept of expertise in risk society, the increasing 'democratization' (ibid., 191) of the notion of expertise is also considered, with those traditionally considered to be lay individuals demonstrating a degree of expertise in their own right.

Given the context of debate and discussion between experts, and the new-found expertise of the audience for this debate, Chapter 5 then considers why this audience might be particularly prone to accepting certain interpretations of the causal chains at work in relation to speed and crash risk, and to rejecting others. The construction and defence of *identity* within risk society is considered through an exploration of the various attempts by participants in the debate to variously adopt and resist the roles of definer, instigator and victim of 'risk' (ibid., 29).

Chapter 6 then considers how the specific methods of implementation used within the NSCP contribute to its perceived illegitimacy. The issue of the production and reception of *justice* within a risk society is therefore the focus of this chapter, which uses literature relating to the concept of procedural justice to explore the reasons for claims of unfairness and injustice in relation to speed limit enforcement.

Chapter 7 then considers modifications and developments that have accompanied the introduction and use of speed cameras as a road safety technology, and also presents some suggestions put forward by drivers themselves with regard to methods of implementation that would be experienced as more fair and just. This chapter then seeks to demonstrate the transferability of the lessons learned from analysis of this particular context, by expanding its focus to other regulatory contexts infused with 'risk' in various forms. In doing so, its intention is to be useful for those engaged in both educational and engineering approaches to road safety, practitioners, law enforcement officers, policy makers at international, national and local levels, as well as (in a different way) to academics concerned with criminological and sociological questions, psychologists and lawyers. Chapter 8, the conclusion, reflects on the findings of the research.

Chapter 2
A Brief History of Speed Limit Enforcement

Given the apparent centrality of the speed camera as a technological device to the debate that forms the subject of this research, the experience of speed limit enforcement *before* such enforcement technologies were used is of interest.

Since well before the introduction of the motor car to the UK's roads, the idea of limiting the harm posed by moving vehicles by limiting the speed at which they could travel had been formalised in law.[1] The use of the criminal law to enforce such limits centralised the responsibility for the vehicle's speed on to its human controller, and such laws were considered no less appropriate for the human controller of a motor car than they were for those responsible for a horse, a steam-powered vehicle or a bicycle. Speed limits have therefore always been used as proxy indicators for the level of risk posed by a driver's choice of speed. Such proxies have been justified on the basis of causal interpretations that connect increasing speeds with increasing risk of harm, fear and, at the turn of the twentieth century, dust (Plowden 1971, 15).

Offences of speeding have therefore always been created by the combination of a law, an act, a limit and a detection instrument, with the limit therefore coming to represent the demarcation of 'safe' (and legal) from 'dangerous' (and illegal) behaviour. Historically, the detection instrument was the human traffic officer, for whom the catching of speeding drivers was just one aspect of a much wider role enforcing traffic law.

When cars first appeared on British roads, the majority of the road-using population (still non-car users) experienced their use not as a pleasurable development but as a risk to their own safety, and there was considerable demand for the enforcement of speed limits against these luxury toys of the rich (Brunner 1928, 9–10). However, for the minority of wealthy individuals who initially benefited from motorised transport, this new-found susceptibility to policing power *as offenders* was considered not only inappropriate but unsettling. Attempts by police constables with stopwatches to catch speeding motorists were quickly followed by motorists' objections to this practice on the grounds that it was not only 'unsporting' of the police (Flower and Wynn-Jones 1981, 42) but distracted them from their 'proper' duties such as catching burglars (Emsley 1993, 361). The police, however, defending their activities in 1904, maintained that they 'set traps only on dangerous stretches of road' where the act of speeding increased the chances of death or injury, and were therefore concerned only with the safety

---

1   The Locomotive Act of 1865 set speed limits of 4 mph in open country and 2 mph in towns (Plowden 1971, 4).

of road users and not with the victimisation of drivers (Plowden 1971, 50). The context in which enforcement was actively pursued was, it seems, considered vital to its legitimacy:

> [Drivers] argued there was no single 'dangerous' speed; it could be entirely safe to drive across Salisbury Plain at 50 mph, dangerous in the extreme to do more than 5 mph down Piccadilly. No general speed limit could cover all cases. Secondly, to insist on a general limit which was patently absurd in some cases could only increase danger. It would be disregarded, perhaps even where it made sense, and this would help to discredit the law as a whole. Laws must be closely related to opinion: if they were not, they were bad laws. (Ibid., 17)

The punishment of behaviour that apparently caused no risk of harm and which was justified by laws with little public support was therefore opposed by those on whose freedom it impinged. Although a minority, drivers were successful in communicating these concerns to the government which, in 1904, passed legislation that made speeding an offence only when it was reckless, negligent or endangered the public. Although intent was still considered irrelevant, the circumstances of the act were considered significant, with the result that 'harmless' speeding was effectively decriminalised. The difference between 'harmful' and 'harmless' speeding was, a result of drivers' protests, recognised in law – laws that regulated traffic for the next 27 years (ibid., 46).

With the advent of cheaper cars in the late 1920s, motoring became a realistic possibility for a much greater number (ibid., 99). From this time on, some sections of the public were increasingly likely therefore to identify with the beleaguered driver, where once they had viewed the debate from the perspective of the endangered pedestrian. This was 'creating friction, hitherto unknown, between the police and the middle classes' (Emsley 1993, 357) to the extent that the police began to see the total decriminalisation of speeding as an attractive proposition (O'Connell 1998, 134). It was, after all, a thankless task:

> In the 1920s above all, the police had resented both having to enforce unpopular laws which involved them in frequent arguments with motorists, and the failure of the courts to back them up. The gibe had been familiar since the earliest days of motoring that while the police hounded innocent motorists, dangerous criminals were left to go about their business unmolested. (Plowden 1971, 393)

Speed limits were abolished, other than in densely populated areas, by the end of that decade, despite rising casualty figures. O'Connell suggests that, having become disillusioned by the concept of the speed limit, 'society sought solace for growing road casualties in the belief that the study of road-accident causes would lead to the design of new roads and propaganda campaigns that would all but eliminate road casualties' (1998, 128). This was supported by a growing stock of research into the problem of motor accidents, aided largely by the collection of data on numbers of road deaths for the first time from 1926 (Foreman-Peck 1987, 267). Such evidence, increasingly used to guide both policy and lobbying by motorists' groups, began to suggest that 'accidents' were not in fact randomly distributed, but had temporal and spatial trends and patterns and involved pedestrians as well as drivers (O'Connell 1998, 127). The education and training of *all* road users thus became a legitimate policy objective in place of the further regulation and restriction of the motorist and the car (ibid., 113).

The 1930s were considered by some to be a golden age of motoring (Plowden 1971, 269). This optimistic vision was, however, short-lived. The year 1934 saw casualty figures of 239,000 (7,343 of which were deaths, although often not of drivers themselves[2]), a figure that 'was not exceeded again in peacetime until 1964 when there were over five times as many vehicles on the roads' (ibid., 271). Despite the objections of the ever-expanding motoring community, the government was convinced: 'however the motorists might protest, their protests should be disregarded in the interests of public safety' (ibid., 280). Finally, in 1934, a speed limit was reintroduced and set at 30 mph, and speeding made an offence for which the driving licence could be endorsed (ibid., 286).

The post-war period is widely seen as being a critical point in the history of the motor car as it was during this time that the car 'finally broke through the class-barrier and on every level became the normal means of transport' (Flower and Wynn-Jones 1981, 166). The influence of the increasing numbers of car owners can also be seen on the legislation and sentencing practice of the period. A Ministry of Transport 'study group' commented in 1960 that:

> Before very long, a majority of the electors in this country will be car-owners. What is more, it its reasonable to suppose that they will be very conscious of their interests as car-owners … It does not need any gift of prophecy to foresee that the governments of the future will be increasingly preoccupied with the wishes of car-owners. (Plowden 1971, 368)

Motorists who killed were consistently let off manslaughter charges, partly because 'juries could not bring themselves to convict a mere motorist of such a "barbarous sounding" crime' (ibid., 345) and no doubt partly because members of the jury could foresee themselves in the position of the defendant (Morris 1966, 31). The 1954 Road Traffic Bill proposed the introduction of a charge of 'causing death by reckless or dangerous driving' specifically to get around this problem and to encourage juries to convict in ways that avoided giving motorists manslaughter convictions (Plowden 1971, 350). Fines for other serious road traffic offences were generally given at only one-tenth of what was allowable, further suggesting that motorists were firmly established as 'not real criminals' (ibid., 345).

---

2   Between 1926 and 1939, pedestrians accounted for around 45 per cent of road deaths (O'Connell 1998, 116).

The 1960s, however, saw an increasing level of realisation about potential problems of mass car ownership, in terms of both the impact on safety and the environment: 'The twentieth century's "love affair with the automobile" has passed through its adolescent passion and settled into a mutual co-existence that often seems increasingly uneasy. There is no question of a separation, of course; just an ordering of priorities' (Flower and Wynn-Jones 1981, 196). This was also the period when, as had happened in the mid 1930s, a 'threshold of tolerance' was reached with regard to road death (Foreman-Peck 1987, 284).[3] The decade saw the introduction of the 'totting-up' procedure which allowed for automatic disqualification for a third conviction for one of a list of over 20 offences (Plowden 1971, 365). An experimental 70 mph limit was also introduced in 1965, becoming permanent soon after (ibid., 370) and credited with a 44 per cent reduction in fatal motorway crashes the following year (Foreman-Peck 1987, 277).

Motorists were able to make their opinions known through an increasing amount of survey research. Fifty-five per cent of respondents to a survey in 1969 admitted to breaking traffic laws (establishing it as, statistically at least, normal behaviour), while fifty per cent also felt that the police spent too much time and effort catching motoring offenders (ibid., 394).

Alongside these developments, an increasing number of authors, drawn from a variety of disciplines, were beginning to write not just about the car as a technological development but about its social, political and criminological significance. 'Victimised' motorists, for example, would have found support for their grievances in the work of J.J. Leeming, an influential author on road crash causation whose work is still cited (see, for example, ABD 2005a). Leeming, a county surveyor in Dorset, objected to the use of the word 'speeding' on the grounds that 'it is a pejorative word tending – and probably intended – to produce prejudice' (1969, 84). He considered that many of the problems and perceived injustice of speed enforcement resulted from its random nature, but that more consistent enforcement was impossible given the amount of offending and the prohibitive cost of employing more police officers to enforce traffic law (ibid., 162). Plowden also suggested that, even if logistically and financially possible, strict enforcement was an unlikely development, given that the police in 1971, just as in 1903, 'dislike[d] enforcing road safety partly because of their discomfort at having to deal with an offender who so often will not "come quietly"' (1971, 393). In agreeing, the writer and journalist Alisdair Aird also provided some evidence of the various methods the police had at their disposal at this time, noting that:

> When the police do make determined efforts to catch these law-breakers, many motorists affect outrage. They imply that there is something fundamentally immoral when the police fall back on the few devices which really succeed in

---

3   The actual road casualty figures represented 'only' a 50 per cent increase since 1938, while the number of vehicles on the roads had tripled over the same period (Plowden 1971, 377).

catching illegal speedsters – plain-clothes cars, radar, air switches on the road surface, and so forth. (1972, 185)

Writing at the same time, Morris used motoring offences as his main example of something that was 'not real crime' to compare with genuine offences such as property and violent offences. He noted that 'the Law Society had suggested that certain kinds of motoring offences, those not involving an element of wickedness but merely stemming from carelessness or poor driving skills, should be removed from the offence category and dealt with outside the socially pejorative atmosphere of the criminal court' (1966, 16–17). This reappraisal was, however, he felt, more to do with the sheer number of offences that still threatened to 'clog up' the courts system than it was to do with 'a systematic reappraisal of judicial philosophy' (ibid., 17). The prevalence of motoring offences was clearly a significant factor in the questioning of their appropriate legal status, but a number of writers also noted that the concept of the speed limit was controversial given that it made possible the criminalisation of behaviour that produced no actual harm. Morris noted that the absence of an identifiable injured victim had contributed to the belief 'that there is really no crime *when no harm can be seen to be done*' (ibid., 20). He later posed the question: 'Who suffers when a man speeds down an empty street? The scope for rationalization is considerable, and it seems likely that many otherwise "honest" citizens take advantage of the fact' (ibid., 22). As such, the prevalence of offending could be contrasted with the 'comparative rarity' with which accidents actually happen to *us* as opposed to *other people* (Whitlock 1971, 7):

> We all make driving errors; and most of us are prepared to take a chance on occasion. In the majority of instances we 'get away with it'. The risk taken is rewarded rather than punished and, inevitably, new learning in the shape of better driving behaviour does not take place. (Ibid., 8)

Such positive reinforcement of law-breaking behaviour contributed to the belief of a high proportion of drivers that they were above average in terms of their driving skill levels – opinions largely held by those who were also particularly intolerant of and aggressive towards other drivers (Aird 1972, 191).

It could also be suggested that the differential treatment reserved for motoring offenders within the criminal justice system (on the occasions where harm did result from the actions) contributed to the idea that it was 'not real crime' but a technical infringement. Motorists were often fined instead of imprisoned, permitted the 'symbolic concession' of being allowed to stand in the well of court or witness box rather than the dock (O'Connell 1998, 134) and only disqualified in extreme circumstances (Morris 1966, 22). Morris attributed this to the fact that:

> there are more people who *could* be in the position of the killer driver – the legions of speeders, light jumpers and careless drivers – drivers on their way

home from 'evenings out' full of good wine – than there are in the position of the labourer who gets involved in a pub fight after too much beer. (Ibid., 31)

This 'ordinariness' of the driver and his (as it was predominantly still a 'he' that featured in writing at this time) presence throughout society was also given by Aird as an explanation for lenient treatment:

> The convicted include those who make laws, those who administer them, those who talk about them and form opinion about them, those who lead and govern … the probability of conviction hangs over most electors. When the convicted criminals outnumber the innocent, is it surprising that the guilty are exceptionally favoured? (1972, 182)

Morris was also able to detect a class dimension to this issue:

> Most of the individuals who are convicted of 'orthodox' crime are lower class. Our society is predominantly middle class and the law-makers and judges are for the most part upper class. There is no problem in condemning the criminal because he is conveniently different from the majority of those who have to evaluate his behaviour and decide on his fate. (Morris 1966, 33)

The enforcement of laws that sought to criminalise the 'wrong' people, simply for their involvement in behaviour which could (but in the vast majority of cases did not) cause danger, was still raising uncomfortable questions for a society that preferred to think of criminals as other people and which saw its relationships with 'its' police from this perspective (Plowden 1971, 414). The 'road hog' was, therefore, a useful construction which allowed offending behaviour to be attributed to other drivers. Such scapegoats reflected class divisions as well as other concerns of the time, and reinforced the minority status (howsoever conceived) of offenders:

> Depending on one's particular tastes, persons on the highway who irritate, impede, or threaten are women, Jews, wogs, psychopaths, or individuals who in the driver's opinion have little right to be on the road and even less to own a car. These stereotypes are invoked and vilified when aggressive feelings are aroused, and it is noteworthy that membership of an 'out-group' is often ascribed to the sources of our annoyance. (Whitlock 1971, 140)

As such, both the legitimacy and the symbolic meaning of the regulation of traffic were increasingly occupying the various experts with a stake in the subject, with the issue of speed limits often forming the focal point for these analyses. This intellectual argument was played out in a context of lowering speed limits, partly brought about by the 1973 oil crisis, but in general persisted with because of their perceived road safety benefits (Plowden and Hillman 1996, 32).

The start of the 1980s saw 20 million vehicles competing on Britain's roads and the introduction of a new points-based system for totting up driving licence endorsements. Fixed penalties for minor offences were introduced in 1986, while the collection of 12 or more points in a three-year period now resulted in a short disqualification for the convicted driver. The first national targets for reducing road casualties by a third by the end of the century were introduced in 1987 (DETR 2000a), to tackle road casualties which at this time hovered around 49,000 killed or seriously injured (of which 3,600 were deaths) per year (National Statistics 2002).

The Use of Automated Detection Technologies

A growing interest in the role of speed in producing these casualties (see Plowden and Hillman 1996, 31) and the recommendation that 'modern camera technology should play a greater role in the context of traffic law enforcement' (Hooke, Knox and Portas 1996, 2) ultimately led to the passing of legislation that permitted the use of speed cameras to detect speed limit infringements (DfT 1991). The evidence of speed cameras was, furthermore, to be sufficient for the prosecution of a driver, meaning that no corroborative evidence from a police officer was to be necessary (Corbett 2003, 21). Various trials of the technology then took place over the next eight years and produced encouraging reductions in both crashes and speeds.[4] It was noted, however, that take-up of speed cameras by the relevant authorities was being limited by the cost implications of installing and operating the technology (Corbett 1995, 352; Hooke et al. 1996, 39).

The government's ten-year road safety strategy, *Tomorrow's Roads: Safer for Everyone*, was launched in March 2000 and set performance indicators in the form of casualty reduction targets. These included a 40 per cent reduction in the number of people being killed or seriously injured in road accidents, to be achieved by 2010.[5] The strategy also contained what was described as 'a shopping list' (PACTS 2000) of ten road safety problems and their proposed solutions, of which one problem was 'speed'. In its conclusion, the report notes that 'speed is our biggest challenge', contributing to an estimated one-third of all crashes (DETR 2000a, 41).[6] This understanding clearly recommends speed as a logical target for achieving the reduction aims set out in the strategy. Within this identified problem, further distinctions were made between excessive speeds (speeds in excess of the posted limit) and inappropriate speeds (speeds within the limit that are nonetheless inappropriate for the conditions). Speed cameras were then mentioned as one way in which enforcement could be used to tackle excessive speeds. Given its position

---

4   These trials are considered in more detail in Chapter 3.

5   Taking the figures of 1994–1998 as its baseline figure.

6   See Chapter 3 for consideration of this statistic and its significance in the ensuing debate.

as *an* option for pursuing *one* aspect of a problem which formed *a* subsection of *one of ten* priority areas, the debate surrounding the use of speed cameras can be seen to have received a disproportionate amount of attention, both within official circles and in the wider public.

In addition to developing its enforcement capacity, the DfT also continued to pursue educational campaigns. The 'speed' element of the DFT's *Think!* campaign, also launched in 2000, centred on the different consequences of driving at 35 mph as opposed to 30 mph.[7] As such, both the education and enforcement approaches being pursued in relation to speed at this time were centred on the speed limit as the differentiation between safe and dangerous behaviour.

The pilot phase of a speed camera-based scheme,[8] intended to last two years and covering eight constabulary areas, was launched in April 2000. The key development contained in the scheme was that it permitted the 'hypothecation' of fine income generated by cameras. By allowing enforcing authorities to retain this income for reinvestment, this development made the use of speed cameras a more realistic prospect for the newly established 'Partnerships' including police, local authorities, magistrates courts, Highways Agency and the emergency services, who had been found to be willing, but not able, to install them (Corbett 1995, 352; Hooke et al. 1996). As such, the pilot programme represented the culmination of the process whereby the identified solution was made a realistic prospect through the introduction of legislation specifically designed to make the use of camera technologies more attractive, given the beliefs about their effectiveness.

An evaluation of the scheme was scheduled for 2002, at which point the desirability of a national roll-out of the scheme was to have been considered. However, preliminary results from the pilot were considered to be so encouraging that the scheme was rolled out nationally after only the first year of operation (DfT 2003a, ii). The first evaluation then took place after the two-year pilot period had elapsed. It provided the figure of a 35 per cent reduction in people killed or seriously injured at camera sites when compared to the long-term trend (ibid., ii). Just over a year after the publication of the pilot report, a three-year evaluation of the (then titled) National Safety Camera Programme (NSCP) was published. A total of 24 police force areas had now joined the programme. This evaluation concluded that there had been a reduction of 33 per cent in the number of collisions resulting in injury at camera sites (Gains et al. 2004, 30). A 40 per cent reduction in the numbers of fatalities occurring at camera sites was also noted, before being converted into the more 'real' figure of 100 fewer deaths (ibid., 1).

---

7   Publicity material included the finding that it takes a further 21 feet, or 6.4 metres, to stop at the faster speed, presented as a fold-out leaflet. Each few metres of this increased distance is illustrated with a picture of a child who would have been knocked down by a car travelling at above the speed limit.

8   The scheme also applied to red-light cameras, but these are not the subjects of this study.

The evaluation of the NSCP then carried out in 2005 found a reduction in deaths at camera sites of 32 per cent, which it also converted into a figure of 100 fewer fatalities per year. The report also demonstrated a positive cost–benefit relationship resulting from the scheme, with 'benefits to society from the avoided injuries' of £258 million for an enforcement outlay of £96 million (Gains et al. 2005, 2). A total of 38 Partnership areas were included in this analysis, covering 41 police force areas, and leaving only two areas not participating in the scheme.

The vision of speed camera enforcement painted in such reports was therefore one of a successful intervention that saved both lives and money. The percentage reductions in casualties evidenced in the evaluation of the NSCP suggest that speed cameras are intervening in the causal chain whereby crashes have resulted in an appropriate manner, and thus that those causal chains were correctly observed and that an appropriate remedy was selected.

While in 1996 only 475 speed camera sites were in operation (Hooke et al. 1996, v), the creation of the NSCP in 2000 saw that figure grow to around 5,000 in 2006 (DfT 2009). In December 2005, the Secretary of State for Transport announced the end of the NSCP and the 'netting off' arrangements that had funded the growth of cameras. As a result of this announcement, from 1 April 2007 Partnerships could no longer recover the cost of enforcement from the revenue it generated. All fine income now goes to the Treasury's consolidated fund in the same way as other fines. The DfT note that '[t]he move gives local authorities, the police and other local partners responsibility for the future deployment and operation of cameras. This allows them greater freedom to enforce in response to community concerns about speeding or at sites where there are speeding problems and a high risk that casualties will occur' (DfT 2009a).

In 2010, the coalition Conservative and Liberal Democrat government announced the end of central funding for speed cameras as part of its cuts to the road safety budget and suggested that local authorities had become over-reliant on the technology. The move was also described as 'another example of this government delivering on its pledge to end the war on the motorist' by the Road Safety Minister Mike Penning (BBC News Online 2010b). Oxfordshire County Council then became the first authority to switch off its speed cameras, with other authorities across the country announcing that they would be reviewing their own use of the technology.

Regardless of these changes to the background funding arrangements, the vast majority of speeding offences are still currently detected by speed camera to the extent that, in 2008,[9] they accounted for 84 per cent of those offences detected (Povey et al. 2010). A range of different camera types have been approved for deployment by the Home Office (see Lewis 2005), but the majority of devices emit infrared or radar beams that trigger the camera if a passing vehicle is travelling at

---

9   The last year for which data are currently available.

a speed higher than the limit.[10] In addition to a photograph showing the speed of a vehicle and allowing for its identification, cameras will also produce some form of secondary check that allows for the speed to be confirmed. One common method of confirming this is the use of white lines on the road, a metre apart, which allow for the speed of the vehicle to be calculated manually.

Following their detection by speed camera, offences of exceeding the speed limit are processed via fixed penalty arrangements. Such arrangements apply to a range of offences contained within Schedule 3 of the Road Traffic Offenders Act 1988. This process is initiated by the sending of a 'conditional offer' to the accused driver via the postal system, containing details of the alleged offence. The speeding driver can then avoid a criminal prosecution in a magistrates' court if, within a specified period of time, they plead guilty to the offence and agree to pay a set fine of £60 and accept three endorsement points on their driving licence. The driver who accepts this 'conditional offer' then sends their licence and a cheque for the amount of the fine to the relevant local fixed penalty payment office, and receives their endorsed licence back in due course. The option of a court appearance is, however, available to those who choose to plead not guilty, and is compulsory for those drivers who already possess nine or more endorsement points. In this latter scenario, the driver is compelled to attend court, where a disqualification for a short period will be considered. As a result, four speeding convictions can result in a driving ban,[11] preventing the driver who has declined to behave in a non-risky fashion from continuing to pose a risk to other road users, or, in the words of the DfT in a recent consultation, 'act[ing] to target in particular those individuals who are guilty of repeated dangerous offences on our roads, and mak[ing] it more difficult for them to offend again' (DfT 2009b, 72).

Between the launch of the NSCP and the end of 2008,[12] over 11.5 million offences of exceeding the prescribed speed limit on a public highway had been detected by speed camera.[13] Between 2005 and 2007,[14] the numbers of drivers *detected* speeding decreased slightly. It is unclear to what extent this reduction can be accounted for by changes in levels of enforcement[15] (Povey et al. 2010) or by increased compliance with limits (DfT 2009b, fig. 7.1), but it seems likely

---

10   Usually the limit, plus a further 10 per cent of the limit plus 2 mph (Association of Chief Police Officers guidelines) to safeguard against discrepancies between speedometers and detection equipment.

11   Some research has suggested that a proportion of drivers with 12 or more points are 'slipping through the system' and avoiding disqualification (Corbett et al. 2008, 55).

12   The last year for which data are currently available.

13   It is not possible to ascertain how many drivers have been detected, given that this figure includes repeat offenders, but this figure is derived from data from Povey et al. (2010).

14   The last year for which data are currently available.

15   Povey et al. in their Home Office Statistical Bulletin acknowledge that the NSCP may have had an impact on enforcement numbers, saying '[t]hese initiatives should be borne in mind when interpreting data relating to the use of cameras' (2010, 51).

that it is a combination of these factors, plus the widespread tactical avoidance of detection, as identified by Corbett et al. (2008) and discussed in Chapter 5.

The use of speed cameras therefore provides a system of surveillance that allows for behaviour designated as 'risky' to be detected on unprecedented levels. Speed cameras have become checkpoints at which people are judged in terms of their ability to behave in a non-risky manner. The surrounding systems of prosecution then provide the means and methods by which risky individuals can be first punished and then, if they continue to behave in this fashion, denied access to the roads. This restriction is therefore designed to reduce the number of risky drivers on the roads through deterring risky behaviour and excluding those who will not be deterred.

Problematising Speed

Offences of 'speeding' are illegal under the Road Traffic Act:

> Under Section 89 of the Road Traffic Regulations Acts 1984 and Schedule 2 of the Road Traffic Offenders Act 1988, it is contrary to the law to exceed the prescribed speed limit on a public highway. (Gains et al. 2004, 14)

It is infringements of this nature that are indicated by the use of the word 'speeding' in this research, although it is acknowledged that 'speed' can be, and is, deemed problematic when it is nonetheless within the speed limit. Where such a distinction is necessary, the two types of problematic speed will be referred to as 'excessive' (speeds in excess of the limit) and 'inappropriate' (a speed within the limit but nonetheless inappropriate for the conditions). This distinction is itself the source of some dispute within the debate.

While they cannot adapt to the changing circumstances that will occur at any specific location, speed limits encourage the use of slower speeds in locations where they are considered to be appropriate, and where higher speeds would be perceived to cause more risk of harm. They also allow for those who decline to limit their speeds in line with these previously conducted risk assessments to be easily and simply problematised.

Speeding is a *mala prohibita* regulatory offence, given that it relates to behaviour that is deemed to increase the probability of harm, but is not pre-existingly 'wrong' or 'immoral' (Duff 2002, 98). The appropriateness of its regulation is therefore dependent on the connection between it and the supposed outcome of engaging in it. Considerable effort has therefore been made to demonstrate the appropriateness of speed limit enforcement through research that links increasing speeds to an increased risk of harm in the form of a road crash.

The minimisation of harm is therefore the motivating concern behind rendering the act of speeding punishable by law and, as such, this punishment can be justified whether or not it was engaged in maliciously, intentionally or even knowingly. The fact that no actual harm need result from an instance of speeding is not relevant because the deterrence of risky behaviour is desirable regardless of whether or not harm actually results from a specific instance. Establishing *mens rea* in relation to the offence is therefore unnecessary given that '[t]he harmful consequence is the same whether it is intended or whether it is not' (Geary 1994, 22) and given the increased probability of harm resulting from the action. By designating speeding offences as strict liability offences, punishable whether or not the behaviour was intended and whether or not harm was actually caused, the 'potential harm-doer' is, theoretically, compelled to exercise greater care to avoid criminalisation by strict liability laws (ibid., 22). Speed cameras are therefore used to encourage drivers to reduce their speeds at locations where speed is considered to pose a potential risk, and excessive speed can be punished whether or not any 'actual' risk was caused.

The recent favouring of speed cameras as the detection instruments that record offences being committed has vastly increased the number of offences that can realistically be detected and criminalised. As a result of the deployment of speed cameras, therefore, motorists are increasingly likely to encounter authority, and the criminal justice system more generally, via automated detection technologies that enforce strict liability legal principles in respect of *mala prohibita* regulatory offences.

Some research has considered the extent of the speeding problem, although little data is available about the demographic profiles of offenders. The prevalence of the offence, however, suggests that it is not confined to any one age, race, gender or class grouping. Although such demographic differences are not the focus of this work, some research has, for example, considered how aspects such as age and gender affect responses to speed cameras. Research has suggested that while more young drivers admit to speeding, they are proportionally less likely to be detected by camera than older drivers (Stradling et al. 2003, 206). The same research suggests that men and women were detected speeding equally frequently by cameras, once adjustments had been made for the average distance travelled by each gender, and were generally exceeding the speed limits by similar amounts when caught (ibid., 209). More recent research, however, suggested that as the use of cameras increased, women have shown a greater propensity to adopt 'conforming' behaviour involving driving at less than the speed limit as a matter of course (Corbett and Caramlau 2006, 419). Men, however, have become increasingly likely to adopt a 'manipulative' approach, slowing down for speed cameras before speeding up again once they have passed them (ibid.). Drivers of cars with bigger engines, high-mileage drivers and younger drivers with passengers their own age have also been found to report more offending behaviour (Stradling 2005, 1–2; Stradling et al. 2003, 96). Generally, more males than females admitted speeding, while the 21–29 age group reported the highest proportion of speeders (ibid., 9). Recent research by Corbett et al. also demonstrates that around half of convicted drivers felt that their speeding had been inadvertent (2008, 1).

However, the interesting statistic from the point of view of this work is the endemic nature of the offence, testified to by consistently high self-reported offending among all groups. Stradling classified 'speeders' as the 37 per cent of the sample who had *been caught* (2005, 2), but it seems unlikely that this accurately represented the speeding population as it excluded those who engaged in the behaviour but had remained undetected doing so. Seventy-nine per cent of Stradling and colleagues' subjects reported breaking speed limits at some time (2003, 5), while Corbett cites research that gives figures of between 85 per cent and 99 per cent (2003, 111). An RAC survey found that 55 per cent of their survey participants were happy to admit to 'exceeding the speed limit a little every day' and were therefore habitual offenders (2005). Such is the endemic nature of speeding that some research excludes those that 'only' admit to doing it 'occasionally' (Corbett 2003, 111). It is hard to imagine such an approach being adopted with other kinds of criminal behaviour – for instance, excluding from research individuals who only 'occasionally' burgled other people's homes. In all cases, speeding behaviour seems to be reported by over half of those questioned and as such can be considered to be an example of 'normal' deviant behaviour.

Chapter 3

Contradictory Expert Claims about Speed Cameras[1]

This chapter and Chapter 4 consider the role of experts and their expertise in the speed limit enforcement debate in the light of Hebenton and Seddon's claim that: '[e]xpertise has never been so indispensable while at the same time so subject to resistance' (2009, 357). The analysis argues that the increasingly 'risk'-concerned context in which the debate about the use of speed cameras takes place has resulted in significant changes to the role of the 'expert', which must be explored if the existence of the debate itself is to be understood, and which must be recognised by those hoping to exert some expert influence in relation to the topic.

These chapters consider, in turn, why experts are needed, who claims expert status and how experts go about constructing their expert case in relation to speed limit enforcement and speed cameras. Their role and expert product is considered in the context of what Beck has termed the 'demonopolization of scientific knowledge claims' (1992, 29) within societies increasingly concerned with issues of risk. This process has, he suggests, led to the co-existence of a variety of competing and conflicting interpretations of the processes whereby risk (in this case, road death and injury) is caused: 'There are always competing and conflicting claims, interests and viewpoints of the various agents of modernity and affected groups, which are forced together in defining risks in the sense of cause and effect, instigator and injured party. There is no expert on risk' (ibid., 29). This context poses a number of challenges to those occupying the position of 'expert', given that the lack of a single definitive 'über-expert' on risk has not heralded the end of the concept of the expert, but has instead changed its meaning and significance. It has allowed a variety of newly enfranchised expert voices to compete for the right to define, and assign responsibility for, risks. This demonopolisation of sources of expertise has, as a result, 'altered the boundaries of officially sanctioned institutions for knowledge production' and dispersed that capability across a wide range of 'expert' sources (Jamieson 1988, 72) such that none can claim to have cornered the market in expertise. Instead, '[d]ifferent purveyors of useful knowledge emerge to compete to supply diagnoses and solutions to practical problems' (Hebenton and Seddon 2009, 353).

---

1   Material from this chapter has also been published as Wells, H. 2011. Risk and expertise in the speed limit enforcement debate: Challenges, adaptations and responses. *Criminology and Criminal Justice* (forthcoming, July 2011).

Chapter 3 considers why experts remain important, before considering *who* the relevant 'agents of modernity' and 'affected groups' (Beck 1992, 29) are in the context of the speed enforcement debate and *how* they go about constructing their expert case in the context of demonopolised expertise.

Analysis then turns, in Chapter 4, to a consideration of how this expertise needs to be *promoted* in order to compete with other contradictory positions. A further consequence of the demonopolisation of expertise is also considered, this time in relation to the impact of the debate upon its audience. It considers that the experts' apparent inability to agree may be partly responsible for a growing independence of the public from such expert sources. As such, the possession of expertise can be said to have been *democratised*. Rather than demonopolisation resulting in an expertise vacuum, a situation is produced in which everyone can, and does, claim a degree of expert status in relation to speed and speed enforcement. That the traditional expert voices considered previously have failed to recognise this fact is, it will be suggested, a further explanation for the debate around speed limit enforcement.

The Need for Experts

The demonopolisation of expertise referred to by Beck is significant because of the importance of the expert as a source of information about technological developments and their side effects. The ordinary member of the public is considered to be ill-equipped to understand the technicalities of the bulk of the problems which confront society:

> As society becomes more complex and knowledge is further compartmentalized into ever smaller arcane specialisms, reliance on experts grows in every feature of daily life … [F]or the most part everybody goes along with the proliferation and increasing influence of scientific expertise, because the benefits of doing so are manifest and, in any case, there is really very little choice in the matter if one wants to be able to negotiate the complex realities of modern life. (Roberts 2002, 255)

Experts are therefore needed both to interpret the processes whereby harm results and to propose interventions to prevent the harm from happening in the future. In recognising a harm and expressing concern about it, therefore, society creates a role for the expert and a reliance on their expert product. However, the fact that many harms are *observable* and that there is general agreement that they are also *undesirable* does not mean that there is an expert consensus as to *what* causes them and *what* should be done to reduce their incidence. Risk issues are, first, issues of chance and probability. They are also, at a variety of levels, matters of interpretation. Because such interpretations cannot claim to be deterministic, disagreements can occur between experts about the accuracy of one interpretation over another. Each interpretation then necessitates an alternative regulatory reality, a different set of implicated instigators of risk and a different type of intervention. Experts are therefore turned against one another and must compete to be accepted as the definitive expert in relation to the issue in question. For experts, the consequence of failing to achieve this definitive status is to experience restriction and regulation justified on the basis of causal interpretations to which they do not subscribe. As a result, a variety of vested interests will be involved in attempting to define the problem in ways favourable to themselves (Beck 1992, 174). Risk debates are therefore public discussions in which various parties are engaged in competing for the right to be definitive in terms of the specific causal interpretations of the processes whereby harms result. Such processes are neither directly observable nor objectively interpreted.

The harm of road death and injury has been identified as a risk issue about which 'something ought to be done' (Hunt 2003, 175). Government has established performance indicators that define what are 'acceptable levels' (Beck 1992, 64) of road casualties and have then embarked on a mission to identify the significant causes of this harm. However, the speed enforcement debate has no shortage of individuals, interests and agencies who claim to be able to ameliorate the effects of mass individual mobility by suggesting *alternative* causal interpretations of the processes whereby harm results on the roads. Road crashes are complex events and causality is not something that, in many cases, can simply be observed from the result. While, to one observer, speed may appear to have caused a collision, to another it may be interpreted only as having exacerbated the consequences of the collision. To a third observer, a wet road may be seen as the cause of a crash, or the choice of inappropriate speed in the conditions may be blamed. This choice may, in turn, result in the cause being seen to be speed, weather or driver error, for example.

As such, there exists a market of competing claims and interests that propose a variety of alternative, and in many cases contradictory, explanations for the undeniable reality of road death and injury, and propose various interventions which they claim will prevent it. Various voices then claim expert status in relation to the particular risk issue by claiming knowledge and experience in excess of the ordinary lay member of the public. Alternative causes, alternative restrictions or alternative interventions can then be proposed which they believe more accurately reflect the reality of road use, or which impact more favourably on them. The state is not therefore the only source of expertise about the causes and effects of risks. It must compete with other sources, and other interests, for the mandate to regulate the behaviours it views as being 'risky' in the traffic context.

At the heart of this particular debate are some critical questions relating to the causal chains implicated in the current policy and the connection between speed and harm. Given the absence of a moral basis underpinning the majority of road traffic regulation, the causal chains linking actions to harms are the only justification for enforcement action taken against those actions (Duff 2002, 97). The legitimacy of targeting speeding drivers is therefore dependent on the proving of a connection between excessive speed and an increased risk of a road crash. Depending on how effectively these links are proved, punishing the action of exceeding the speed limit can then be seen either as an entirely warranted regulation that targets instances of violent manslaughter or as 'the malicious exercise of arbitrary power' (ibid., 99) through the entirely unjustified punishment of a behaviour that has no harmful consequences. If the latter view is held by those over whom this power is being exercised, their compliance, cooperation and consent are unlikely to be forthcoming. Demonstrating and promoting these links, and negotiating public acceptance of them, is therefore as much a priority for those who wish to pursue enforcement on this basis as establishing those connections in the first place.

The Construction of Expertise in Risk Society

The emphasis on difference and dispute contained within the notion of the debate around speed limit enforcement should not obscure the fact that the debaters share certain characteristics and use certain methods for identifying and communicating the causal interpretations they favour – for claiming expert status. Before considering the various groups and agents claiming expert status within this debate, it is therefore necessary to consider on what grounds expert status is claimed.

Instead of a single omnipotent expert on risks, a variety of self-proclaimed experts compete for the right to be definitive. As a result, experts are seen to occupy a strange and contradictory position. Although there remains an underlying faith in the concept of the expert itself, this faith is continually rocked and challenged by the existence of alternative and competing expert voices striving to undermine each other. The possession and promotion of *evidence* of the interpretation they favour is the only way experts can set themselves apart from one another and distinguish their 'expertise' from simple 'opinion' or vested interest. The construction of such an evidence base is therefore a priority for a debating expert. While there may be no such thing as a monopoly on expertise in risk society, scientific expertise has the 'monopoly on truth' (Beck 1992, 71) and is thus an integral part of an evidence base. Supposedly, everything is subject to redefinition, to debate, '*except scientific rationality itself*' (ibid., 164, emphasis in original), with the result that the terms of engagement for the debate are agreed upon, despite the lack of a consensus as to what these scientific principles conclude. Challenges to one set of scientific interpretations must therefore be phrased in the same scientific language if they are to draw expert status from this same set of criteria, and the resulting debate is consequently, on one level, an exchange of competing scientific interpretations:

> [B]ecause risks are brought into existence by scientific knowledge and law, they can also be magnified, minimized, or muzzled by scientific knowledge and law. Those who dispute how standards are magnified, minimized, or muzzled by scientific knowledge and law are always required to answer with greater scientific knowledge. (Ericson and Haggerty 1997, 98)

The following analysis therefore considers each significant expert voice with a presence in the speed debate, alongside the expert case that these voices promote, and the scientific basis on which they stake their claim to definitive expert status in relation to speed limit enforcement.

Who Claims Expert Status in the Speed Camera Debate?

A variety of authorities, agencies and individuals have claimed expert status in relation to speed and speed limit enforcement. These are the significant voices that are heard whenever the subject of speed-related interventions into road safety contexts are debated in the media, among other experts and among the public. This combination of traditional and newly enfranchised expert voices represents a variety of different interests and motivations which vary in size, strength and influence. The voices represented here were selected through analysis of media reporting of the debate and through a process of inter-referral, whereby experts' references to other significant experts they encountered were used to generate a list around which this research could be focused.[2]

Each expert is considered in terms of their 'debate profile' and the 'case' that they put forward for consideration within the debate and which constitutes their argument for a particular causal interpretation. Each section considers a particular expert's profile in terms of its media presence, other output (press releases, speeches, strategies, internet presence, etc.) and references made to it by other experts within the debate. Each expert case consists of the evidence base on which they draw in the construction of their input into the debate and the scientific research, analysis and evaluation that they use to reinforce and justify their stance in relation to speed limit enforcement. First, the expert case set out by the state in favour of speed cameras is considered, before other voices (both 'for' and 'against') are explored.

The Official Picture: The State's Case for Speed Cameras

In any risk debate, certain authorities are in a position of power which allows their interpretations of the causal chains at work within a risk issue to come to dictate and shape the regulatory reality that relates to it. In many cases these will be state authorities that determine both that something should be done about a problem and what that intervention should be. In doing so, they acknowledge the existence of an injured party and create an implicated instigator of the risky behaviour which their interpretation has identified. In the context of demonopolised expertise, however, these interpretations must compete with other interpretations put forward by other experts. Such experts do not necessarily have to act to

---

2   See Chapter 1 for a full explanation of these selection processes.

establish their expert status, but must act to *maintain* their expert credentials and to *defend* their status. As such, the demonopolisation of expertise can be seen to have particular significance for those agencies that might previously have been thought to possess a monopoly on policy-relevant expertise, and which have had their monopoly overturned by various newly enfranchised challengers. As Hunt notes: 'Experts are no longer satisfactorily legitimated by their location in state institutions or professional organisations. They are increasingly dispersed across rival institutional sites and disseminate a plethora of incompatible knowledge and advice' (2003, 169). As a consequence of this dispersal of their product across rival suppliers, 'status' experts of this type cannot assume that their particular brand of expertise will be automatically accepted, or that the restrictions they wish to enforce will be universally welcomed. Previously privileged status experts of this kind are compelled to legitimate their interpretations by emphasising the evidential basis on which they are formed. In doing so, they make a case for the superior interpretative accuracy of their own interpretations, over and above that of alternatives.

This section therefore considers who the relevant status experts are in the speed enforcement context, the sources from which they seek to obtain the research that forms their evidence base, and the outlets that they use for communicating this evidence.

The Home Office and Department for Transport (DfT) were the two central government agencies with an interest in, and responsibility for, the enforcement of speed limits when speed cameras were first introduced to UK roads. While the DfT had lead responsibility for managing the National Safety Camera Programme (NSCP), the use of the criminal law to underpin enforcement created a role for the Home Office.[3] These central government agencies were the privileged regulators in a position to bring about real consequences through the formulation of policies that operationalised the causal interpretations to which they subscribed. They were, and the DfT and Ministry of Justice now are, responsible for crucial elements of the speed limit enforcement policy, such as the setting of speed limits, and for establishing the means and processes whereby punishment for breaching them can result. They are also concerned with creating policies that are acceptable to those who will be regulated and controlled as a result of them. Their expertise is presented in the form of evidenced policies and strategies that are fed into the debate, where they are defended in response to challenges from other experts. In the context of demonopolised expertise, these government agencies have therefore been faced with the task of attempting to maintain their credibility and their mandate for regulating behaviour.

As both a criminal justice and road safety policy, the NSCP was considered to be unique in that it had required a level of justification never before necessitated by a policy designed to reduce harm through the punishment of law-breaking

---

3   This responsibility passed to the Ministry of Justice following the reorganisations of 2007.

behaviour. As one of the consultants involved in implementing the scheme noted, road safety policies had 'never caused so much debate': they had 'normally been pretty uncontroversial vote-winners' (consultant employed by DfT).

The police are centrally implicated in carrying out the enforcement required as a result of the causal interpretations favoured by the DfT and Home Office. The police have a significant historical function as definers of society's problems (Loader and Mulcahy 2003), and their deployment in the pursuit of risky behaviour (as opposed to morally and criminally proscribed behaviour) continues this defining role in a particularly current way. However, having been seen to have been rendered 'profane' to a degree (ibid., 32–3), the police are also reliant on the scientific legitimisation of the policies they enforce. They, too, require a convincing evidence base to demonstrate that their deployment in relation to the issue of speed is legitimate. Such an evidence base is either borrowed from central government policy sources or produced through their own experience of enforcement (see, for example, ACPO, DfT and Home Office 2005, 4; ACPO 2004, 4).

The approach to road safety taken in the policy under discussion here has also created a role for a variety of agencies and authorities whose activities have a direct impact on some aspect of the causal interpretation favoured by the government. The involvement of local government in setting speed limits, the Highways Agency in managing motorways and trunk roads, the police in providing an enforcement capability, fire and rescue services in attending crash scenes and the magistrates' courts service in processing and punishing offenders means that each of these groups is drawn into the debate. The formation of 'Partnerships' of these agencies to enable coordinated implementation of the policy has invested all of the partner agencies with a new degree of expert status in relation to this issue. Created in legislation of 2001, the Partnership model simultaneously distances the policy from a single, individual expert and draws in others to create a new enforcing body in its own right, variously known as a Safety Camera Partnership, Speed Camera Partnership or Casualty Reduction Partnership. These bodies are able to enforce their own, and the government's, interpretations of the causal chains involved in speed-related crashes through their responsibility for all aspects of camera siting, installation, operation and maintenance, as well as detection, prosecution and punishment, education and publicity. This Partnership method has also served to bring together all the state's official experts to give the impression of a consensus, attempting to combine status experts, experts with a traditional aura (Loader and Mulcahy 2003, 33) and scientific experts to create a repository for expertise and to engineer definitive expert status under the Partnership banner. Partnership websites often publicise the locations of cameras, research findings and government targets, as well as some form of 'frequently asked questions' section where some of the generic and familiar criticisms of the policy are acknowledged and addressed.[4]

---

4   See, for example, www.gosafe.org/default.aspx, www.cprsp.gov.uk/public_faq. aspx and www.staffssaferroads.co.uk.

As a newly enfranchised expert voice, Partnerships have had to compete with the voices of their more familiar constituent agencies for a presence in the debate. The high visibility and symbolic meaning of the police, for example, means that their voice is more often sought and reflected in media reporting in this area. Partnership arrangements may also appear less accessible to the media, who prefer to represent the opinions of a more familiar, and thus meaningful, expert. For example, a Partnership as a body was mentioned in 47 of the 176 press articles surveyed as part of this research, while Partnership agencies were mentioned *as individual bodies* (despite their Partnership membership) in 50.

The residual advantages of status mean that experts of the type profiled above have little difficulty securing outlets for the promotion of their causal interpretations, through their own publications, strategies, research and press releases and, importantly, through media channels. Within the newspaper articles analysed as part of this research, 250 references were made to a status expert of this type,[5] that being the DfT, Home Office, police or Safety Camera/Casualty Reduction Partnership. If each article was to be viewed as a microcosm of the debate, however, these voices were effectively 'unchallenged' in only 80 cases; in the remaining 61 articles, another competing expert source was also referenced and therefore effectively broke the status expert's monopoly on expertise in that article, presenting the topic of speed to the audience as a contested issue.

Before continuing to profile the other sources of expertise present within the speed limit enforcement debate, the evidence base on which status experts rely for the justification of their interventions will first be considered. This evidence base will be considered in some depth, given that the task of defending their chosen intervention and attempting to negotiate acceptance of it are tasks that are the particular responsibility of status experts, and which therefore represent a challenge to these agencies specifically. Their interventions and causal interpretations are, it is suggested, particularly under pressure in the context of demonopolised expertise and the subject of intense scrutiny. They are tested on a daily basis in the full glare of the public and media gaze, and are held accountable annually when road casualty figures are released and taken as evidence of the accuracy or inaccuracy with which causal chains have been observed and intervened in.

Proponents of speed cameras are able to draw on a stock of research to support their case for the use of speed cameras, research on which they base their claim to expert status. This has not always existed, however, and the two decades since the arrival of speed cameras on British roads have seen a variety of studies and evaluations of their impact in terms of speed reduction, crash reduction and public acceptance.

The official stance on speed limit enforcement maintains that harm in the form of road death and injury can be lessened by reducing speeds through the use of technologies that encourage compliance with speed limits at specific points. A priority has therefore been to source or produce research that demonstrates that

---

5   A total of 141 articles referenced one or more status expert.

speed causes crashes, that reduced speeds reduce crashes, that cameras reduce speeds and crashes, and that this form of control is accepted along with the rationale that underpins it. As such, a large number of research studies, evaluations and surveys have been used to justify the use of speed cameras and to demonstrate that the method has public support. As the use of the technology progresses, the emphasis on different findings and evidence has also shifted to include the most recent and directly relevant research available. The appropriated and commissioned research evidence is explored, below, in terms of its contribution to the creation of a convincing case in favour of the use of speed cameras.

Following the recommendation of the 1988 *Road Traffic Law Review Report* (North 1988) that 'modern camera technology should play a greater role in the context of traffic law enforcement' (Hooke et al. 1996, 3), the use of speed cameras as an enforcement tool was permitted in the 1991 Road Traffic Act. Plowden and Hillman suggest that their own research into the effects of speed limits in over 30 countries also led to speed reduction becoming a political priority (1996, 31–2). The deployment of speed cameras to enforce speed limits and thus to reduce accidents was a logical outcome of these three developments.

The first research into the effectiveness of speed cameras at reducing speeds and crashes followed in 1992. This was effectively the first applied test of the assumptions about camera effectiveness and the effects of reductions in speeds. The study compared the speed of traffic and number of crashes on roads in West London for six months before and six months after the installation of speed cameras (London Accident Analysis Unit 1997/2003). This research concluded that cameras were effective at reducing both speeds and crashes, and has been used subsequently, and as recently as 2005, as part of the evidence base in support of the use of speed cameras (see Corbett and Simon 1999; DfT 2005a).

In 1996 the Home Office Police Research Group published further research that considered the costs and benefits of traffic cameras, broadly conceived. Potential costs and benefits related to economic matters and the cost-effectiveness of operating cameras, the influence on crash frequency and, crucially, any effect of the use of automated enforcement on public opinion (Hooke et al. 1996, v). The inclusion of this latter aspect as a potential cost or benefit suggests that, even at this early stage, the acceptability of cameras was considered to be as important as their effectiveness in producing a policy that could be considered 'successful'. The paper noted that 'research has demonstrated that securing greater compliance with road traffic law can play a key part in reducing accidents' (ibid., 1–2), before indicating that '[r]esearch by the Transport Research Laboratory (TRL) suggested that each one mile per hour reduction in average speed can produce a 5 per cent reduction in injury accidents' (ibid., 2). By linking the two research findings, it is thus able to conclude, along with an earlier DfT circular, that '[i]ncreasing compliance with speed limits and traffic light signals should therefore bring about significant reductions in accidents, thereby contributing to the overall target for casualty reduction' (DfT circular 1/92, cited in Hooke et al. 1996, 2).

Research is therefore used to promote the assumption that reduced speeds produce reduced crash incidence, and that promoting compliance with the speed limit is an effective means of reducing speeds. It is then assumed that speed camera technology is one means by which this compliance can be encouraged, but, at the time of the circular quoted above, no specific research that tested these assumptions could be called upon.[6] This absence was, in fact, given as part of the reason for the commissioning of the Police Research Group research, which could then fill this gap in the official evidence base in support of speed cameras (Hooke at al. 1996, 3).

The 1996 study used data relating to a total of 420 speed camera sites (ibid., vii). It demonstrated that speed cameras generated significant amounts of fine income, that they reduced both accidents and speeds (by 28 per cent and 4.2 mph respectively), and that there was no evidence of 'public hostility' to the new technology (ibid., 22). No evidence was found to suggest that crash reductions were caused by factors other than the introduction of speed cameras.

In 1998 a Transport Research Laboratory publication (TRL 323) produced evidence from which further claims about the role of speed in crash causation were made. The study itself considered a new system of recording contributory factors to road accidents, focusing on the recording system itself, rather than the causal factors it revealed (Broughton, Markey and Rowe 1998), but this data was extrapolated to produce figures directly applicable to the evidence base in favour of speed restrictions. The claim that the study demonstrated that speed contributed to one-third of all crashes has become one of the most contested aspects of the debate and, as such, forms a case study in itself. This is considered later in this chapter; for now, it is sufficient to note that this 'one-third' figure became a key weapon in the official case for the use of speed cameras in the policies, strategies and reports that followed its publication in 1998.

As the specific origin of much of the road safety activity that has emerged over the last six years, the use of evidence in the DETR's *Tomorrow's Roads: Safer for Everyone* road safety strategy, published in 2000, is of particular interest. It was in this document that the emerging faith in a combination of technology and speed limits evidenced in previous research came to fruition in the form of a national policy. The strategy notes that '[c]ameras have proved their effectiveness in enforcing speed limits and reducing speed-related accidents and casualties at accident hot-spots', but does not specify the research that has allowed it to reach this conclusion (DETR 2000a, 61). The strategy does, however, allude to TRL 323 by advising that 'research has shown us that speed is a major contributory factor in about one-third of all road accidents. This means that each year excessive and inappropriate speed helps to kill around 1,200 people and to injure over 100,000 more' (ibid., 37). As an important strategic document, this publication is subsequently cited by ensuing pieces of legislation as the source of the claim that speed is implicated in one-third of all road crashes, despite not giving details

6   Research into the 1992 West London project was not published until 1997.

of what this evidence is or where it can be found. The adoption of this figure of 'one-third', however, ultimately leads to the conclusion that 'speed is our biggest challenge' (ibid., 41) and clearly recommends it as a logical target for achieving the ambitious casualty reduction targets set out in the strategy. At this point, a distinction is made between excessive speed (of the kind that might be expected to be influenced by the use of speed cameras) and inappropriate speeds (which are potentially dangerous despite being within the limit in force at a particular location). The emphasis on enforcement contained within subsequent parts of the strategy, however, focuses only on the former aspect of 'speed'. The potential for speed cameras to make a significant contribution to the targets set out in the strategy is clearly a motivation behind the strategy's promise to introduce a new funding arrangement for speed cameras. It notes that, rather than fine revenue being collected centrally by the Treasury, 'the Government now accepts that those responsible for installing and operating cameras should retain some of the fine revenue from offences detected by cameras, to cover their costs' and promises that this system would be in effect from April 2000 (ibid., 61). In doing so, the strategy signals the start of the 'hypothecation' of fine income, which made the subsequent NSCP a realistic, and indeed attractive, proposition for the agencies involved in it.

Also in 2000, the DETR published a review of policy in relation to speed entitled *New Directions in Speed Management* (DETR 2000b). The review's methodology involved consulting a wide range of experts, including 'representatives of environmental interests, the police, academics and many others to reach an informed view of the issue' of speed and with a view to making recommendations for improving speed-related policy and practice (ibid., 3). In terms of its evidential base, the review is more explicit about the research literature it uses than the *Tomorrow's Roads* strategy (launched the same month). However, it occasionally confuses the issues of inappropriate and excessive speeds. Four of the seven research studies included in the appendix relate to speed issues other than that of speeds in excess of speed limits, yet the conclusion that '[t]he most pressing concern is to make drivers comply with existing speed limits' is still reached (ibid., 23). This is also despite the acknowledgement that '[a]t present the problems on rural roads mostly concern vehicle speeds that are within the current limit but not appropriate for the conditions' (ibid., 24). In addition to the 'overwhelming evidence' from 'national and international literature' (ibid., 11) about the connection between increased speeds and the increased risk of a collision, the report also credits more commonsense interpretations. These are again based around TRL 323:

> Broughton et al.'s (1998) [TRL 323] work indicates that *excessive speed* was a contributory factor in 424 of the 2795 accidents studied (about 15 per cent). But this is likely to be an underestimate. *Speed* will have been a part of the reason for other factors such as failure to judge another person's path or speed, which caused 623 of the accidents, about 22 per cent. It is not possible to quantify these contributions directly. (Ibid., 12; emphasis added)

The figure of 'around a third' is therefore obtained by combining the figures for excessive speed and 'other' speed. The strategy's later recommendation that enforcement of speed limits be pursued in order to achieve road safety aims is therefore a recommendation for targeting the 15 per cent directly attributed to excessive speed. Despite this, later publications relating to the NSCP continue to refer to the figure of 'one-third' as justification for both the initial existence of, and the expansion of, the programme (see Secretary of State for Transport 2002).

It was alongside these publications, reviews and strategies that the pilot phase of the NSCP was launched in 2000. As such, they represent the stock of expertise that justified the use of speed cameras as a road safety intervention at this initial stage. The figure of 'one-third' features prominently in the evidence base being deployed at this time, and the accompanying press release notes that '[t]his is all about safety. Research shows time and time again that excessive speed causes injuries and costs lives … We hope that the new funding mechanism being piloted in these schemes will ultimately bring about a reduction in accidents at the sites where the cameras are deployed' (DETR 1999a).

In 2002 the Transport, Local Government and the Regions (TLGR) Committee published its critique of government policy in relation to road traffic speed and recommended areas in which improvements could be made (TLGR Committee 2002). The report's conclusions were based on over 150 submissions from those with an opinion on road safety, including sources as diverse as the BBC, the Association of Chief Police Officers and the 'Sutton Seniors Forum'. As such, this research focuses on expert testimonies rather than evaluation reports of technology in use, but nonetheless presents such views as a sound basis on which to make policy. Its contribution is to present an expert consensus on the issue of road traffic speed and it notes that only two out of 157 expert submissions disagreed with its views on the importance of speed as an enforcement priority (ibid., 11).[7] The committee's views on speed are summed up in the following statement: 'The largest single contributor to casualties on our roads is driving either at excessive (breaking the speed limit and therefore illegal) or inappropriate (i.e. speeds which are foolish for the conditions even if within the speed limit) speeds' (ibid., 7). The source given for this statement is the *Tomorrow's Roads* strategy which, as already noted, is not referenced itself. However, the report does include substantial research and other evidence linking speed (generally) to accident severity and to the likelihood of accidents occurring in the first place (ibid., 10). This research is mainly drawn from academic and local research projects. Its faith in the trustworthiness of research evidence is demonstrated in the following statement: 'We know what to do to reduce the casualties. The Government has commissioned research and funded the pilot projects which show what should be done' (ibid., 64).

---

7   The RAC Foundation and Association of British Drivers (see below).

The report concludes by highlighting the importance of the issue of evidence, and in so doing draws attention to the balance that the government had to maintain between hard evidence and acceptability:

> The Prime Minister has recently rightly stressed the importance of basing decisions on scientific analysis. He now has to decide whether Government policy on speed will be dominated by concerns about how it is portrayed by a section of the motoring lobby and in parts of the press. The alternative is to base it on the detailed research of experts, including TRL, the AA, and the Royal College of Physicians. (Ibid., 65)

The passage leaves the reader in no doubt as to which scenario the committee prefers and considers to be the appropriate way to formulate policy, but it also alludes to the growing controversy about the use of speed cameras and the contrasting views of other voices within the debate.

In 2003 the Department for Transport then published *A Cost Recovery System for Speed and Red-Light Cameras* (DfT 2003a) which evaluated the first two years of the pilot programme, launched in 2000 and covering eight police force areas. From this point on, proponents of camera enforcement were able to supply evidence of their activity that was directly related to the policy in operation at that time. Significantly, TRL 323, which received such prominence when the NSCP was launched, is omitted from the evidence base acknowledged at the start of the evaluation of it only two years later. The evidence base justifying the launch of the NSCP pilot was, instead, given as 'a large number of research studies ... both in the UK and abroad' which testified to the success of cameras, while the link with speed is also based on 'a large number of research studies' (DfT 2003a, ii). Only one reference is given for each case: the 1996 Police Research Group report (Hooke et al. 1996) for the success of cameras, and another Transport Research Laboratory study, TRL 421, for the significance of speed in causing crashes (Taylor et al. 2000).

The pilot areas used a variety of fixed-site, mobile and digital camera technologies so that the pilot could identify best practice (DfT 2003a, iii). Speed surveys were conducted before and after the installation of cameras to determine if the technology had reduced the speed of passing traffic, while a complex comparison of crash and injury data was also conducted.[8] The study provided the figure of a 35 per cent reduction in people killed or seriously injured at camera sites, when compared to the long-term trend (ibid., ii). This conclusion was reached after the completion of the pilot period of two years, but the government had considered the scheme to be so successful that the programme had been rolled out after only the first year (ibid., ii). Despite this, the report claimed that '[t]

---

8   This included comparisons with national trends and some adjustment of figures to take account of seasonal variations in crash rates. For further methodological information, see DfT 2003a, 4-1.

here is strong evidence that these reductions [in speeds and in crashes] have been sustained over time' (ibid., iii) and that public opinion remained supportive (ibid., iv). However, the observation was also made that 'areas focused predominantly on existing sites performed less well compared to areas that introduced new cameras' (ibid., 4-6), suggesting that one year might not have been sufficient to assess whether any tailing-off in effect was in evidence. The impact of increased numbers of cameras on public opinion could also not be predicted and, as subsequent events showed, should not be discounted as a significant influence on the policy and its effectiveness.

From the outset of the evaluation of the NSCP, the three main conclusions highlighted were that speeds had been reduced, crashes had been reduced and 'public reaction has been positive' (ibid., ii). The requirement that Partnerships conduct public attitude surveys was established in the pilot phase, and the 2003 report notes that there was no change in the percentage of the population who felt that 'cameras are an easy way of making money out of motorists' during the course of the pilot. This percentage remained at a (seemingly quite significant) 45 per cent throughout (ibid., 5-3). Generally, however, the surveys conducted by the Partnerships at this pilot stage do seem to evidence a high level of support from the general public. Sixty-eight per cent of survey respondents agreed that '[c]ameras mean that dangerous drivers are now likely to get caught' (ibid., 5-3), while 72 per cent agreed that '[f]ewer accidents are likely to happen on roads where cameras are installed' (ibid., 5-2). This evidence was, of course, also scientifically derived from polls and surveys. The dedication of 39 pages (over half the report) to explanations of the terms, methodologies and analysis on which the report's conclusions are based seems to anticipate the intense scrutiny that these conclusions would be subject to.

One year later, a second report was published which evaluated the first three years of the NSCP (Gains et al. 2004). On this occasion, however, the attributed authorship was a consultancy company and a team of academics from University College London, rather than the DfT itself. Given the increasing availability of directly relevant research now available, this evaluation was able to claim that, in addition to a 'large number of research studies both in the UK and abroad', there existed a stock of directly relevant evidence based on the previous research into the scheme as well as the 1996 Police Research Group paper (ibid., 3). Studies evidencing the role of camera technology in securing adherence to limits and the role of speed in accident causation were dropped in favour of research that evidenced the successful application of the technology itself. As such, rather than *propose* that a crash reduction outcome is the *likely* result of the deployment of the camera as an intervention, this later report is able to provide *actual* evidence of this intervention in action.

The second report considered the experience of the 24 areas now participating in the programme. A methodology similar to that used to the first evaluation report was used, with reductions in speeds at established and new camera sites also compared. It was noted that a straight comparison of 'before' and 'after' casualty data was not possible, given seasonal variations, national trends and other changes at camera sites, such as changes to speed limits or road layout. However, a statistical model was developed that allowed the authors to conclude that a 33 per cent reduction in the number of collisions resulting in injury had been achieved (Gains et al. 2004, 30). A 40 per cent reduction in the numbers of fatalities occurring at casualty sites was also noted, before being converted into the more 'real' figure of 100 fewer deaths (ibid., 2).[9] Opinion polls were again used to illustrate the high levels of public support in evidence in the populace and to challenge perceptions of the scheme's unpopularity. The study found that '79 per cent of people questioned agree[d] with the statement that "the use of safety cameras should be supported as a method of reducing casualties"', although slightly fewer (68 per cent) of those surveyed agreed that cameras *were* primarily used for this purpose (ibid., 7).

The evaluation of the NSCP then carried out in 2005 was again conducted by independent consultants and academics (Gains et al. 2005). The research evidence used in support of the scheme at this stage is able to refer to both previous evaluations.[10] The report notes that the evaluations have considered a steadily increasing number of speed camera sites over an increasing time period, and as such generates an impression of an evidence base growing in both applicability and strength. Again, older and less directly applicable research has been replaced with that which considers the specific intervention of the speed camera under the specific conditions of the NSCP (ibid., 4).

The report found a reduction of deaths at camera sites of 32 per cent, which was again converted into a figure of 100 fewer fatalities per year (ibid., 2). Averages of all the public opinion surveys conducted by participating Partnerships found reductions in the numbers of respondents believing that 'cameras are meant to encourage drivers to stick to the limits' and 'cameras mean that dangerous drivers are more likely to get caught'.[11] An increase in the numbers of respondents believing that 'cameras are an easy way of making money out of motorists' was also detected, rising from 45 per cent in the baseline research (Corbett and Simon 1999) to 55 per cent in this report (Gains et al. 2005, 61–74). The report is able to conclude that '[i]n terms of speed and casualty reduction, public acceptability and funding arrangements … the programme has met its four main objectives' (ibid., 8). The press release accompanying the release of the report noted that '[t]his report is clear proof that safety cameras save lives. There are hundreds of people alive today who would otherwise be dead. All the academics involved in this

---

9   For detailed methodology and results, see Gains et al. 2004, 21–8.

10   Although the Police Research Group paper (Hooke et al. 1996) is referenced in a footnote, this appears to be an error as it is the three-year evaluation (Gains et al. 2004) that is referred to in the text.

11   From 83 per cent in the baseline research (Corbett and Simon 1999) to 75 per cent in relation to the first statement, and from 78 per cent to 60 per cent in relation to the second statement.

independent report agree that all the cameras are delivering substantial reductions in accidents and casualties' (Alistair Darling, quoted in DfT 2005b). The strength of the evidence base and the academic reinforcement of the government's causal interpretations were therefore the focus of this statement. The speed camera's role in saving lives is stated emphatically and evidenced by the stock of research available as the NSCP entered its sixth year of operation.

A 2007 circular, issued shortly before the announcement of the ending of the NSCP in April of that year, refers to the evidence base in support of the use of speed cameras as being comprised of studies including, but not limited to, the Hooke et al. (1996) report, the 2003 pilot evaluation and the two subsequent NSCP evaluations (2004, 2005). It also notes that all the evidence is readily available on the DfT website. Significantly, in noting the overall conclusions of those reports, it acknowledges the issue of 'regression to the mean', noting that 'even after allowing for this phenomenon, safety cameras still achieve substantial and valuable reductions in collisions and casualties' (DfT 2007, 2). The inclusion of this clarification is evidence of one of the most hotly contested aspects of the debate. The suggestion has been made by the usual critics (see ABD 2001; Smith 2004a) that apparent reductions in incidents at speed camera sites can be accounted for by the phenomenon of regression to the mean, whereby a supposedly freak incident (such as a road crash) occurs randomly in one year but does not occur in the following year. Critics suggest that the installation of a camera in the intervening period is not therefore responsible for the reduction, which they suggest is accounted for simply by chance and the fact that 'accidents' are random events. The same three reports (referred to as the 'Two Year', 'Three Year' and 'Four Year' reports) are also offered as evidence on the DfT's website under the 'frequently asked questions' banner for the NSCP.[12]

Other more recent publications referring to the evidence base on which the DfT is relying in operating speed cameras include those used to support a new television campaign. A press release issued to launch the 'Live with it' campaign notes: 'Speed kills. More than 700 people were killed in 2007 in accidents where someone was driving too fast – that's two people every day of the year who didn't go home to their families' (DfT 2009c). This suggests a contribution of speed (although it is not clear whether 'too fast' refers to excessive or inappropriate speed) to around 24 per cent of fatal crashes, although the figure of 700 people makes that statistic more 'real'. The campaign does not make explicit references to speed cameras, and hence evidence of their effectiveness is not relevant; however, the limit (and the 30 mph limit specifically) is still central to the message of the campaign. When the campaign was relaunched in 2010, the figures killed had been updated to 600 to reflect the most recently available evidence (DfT 2010).

---

12   This content has been archived but can still be found at: http://webarchive. nationalarchives.gov.uk/+/http://www.dft.gov.uk/pgr/roadsafety/speedmanagement/safetyc amerasfrequentlyasked4603?page=1

The Road Safety Strategy Consultation, launched in 2009 and covering the period 2010–2020, retains a focus on speed limits, noting that some limits need to be lowered given that compliance with them is good but crashes continue to occur. More generally, improving compliance is identified as an issue, and again the contribution of speed to the total road deaths is noted: 'Of 2,946 road deaths in 2007 there were 727 deaths where speed was recorded as a contributory factor, while a 2007 survey for the Think! campaign showed that over 70 per cent of drivers admit to speeding' (DfT 2009b, 30). The two issues are once again conflated, although the extent to which excessive speeding is responsible for the first statistic is not clear. Speed cameras feature in the document only once, in relation to average speed technologies, in which the priority aim is to publicise evaluation reports of their effectiveness (ibid., 76). However, the role of speed cameras is implicit in references to enforcement of speed limits throughout, in which, under the previous Labour government, they seem to have secured an accepted role in the road safety armoury. Unlike its predecessor ten years earlier, the evidence base in support of fixed cameras is not offered as their role appears to be considered a given.

While not intended to cast doubt on the fact that speed (either excessive or inappropriate) contributes to a great number of deaths and injuries on our roads each year, the purpose of this analysis has been to trace the process whereby this has been created as a *social fact* by government through the use of research and their own deductions. Proponents of the use of speed cameras are now able to quote a variety of research sources in support of speed cameras and the system of deployment currently in use. This analysis has also shown that communicating this causal link and attempting to persuade the public that this link is widely accepted by their peers and by other experts has been a priority for those in favour of speed camera use.

Dissenting Police Voices

While, as has been noted, some agencies and organisations carry with them a certain amount of status by virtue of their history and authoritative position, in the new expert context created by a concern with risk, dissenting voices could be heard *within* this category. These voices proposed and promoted alternative causal interpretations to that of the majority of official expert voices. These alternative voices are, nonetheless, still categorised as status experts, given that they are still able to draw on the same traditional sources of legitimisation, despite their contradictory stance.

A number of significant individuals who are able to claim the benefits of status have adopted an alternative position which opposes the use of speed camera and challenges the evidence base of the experts considered in the previous section. In doing so, these experts have proposed their own causal interpretations of the events leading to the undeniable reality of road death and injury, and have proposed their own interventions which they claim will reduce the incidence of it. The police service, for example, has not always spoken with a single voice in this debate. Its voice has contained a variety of different interpretations, manifested in different policy approaches in different areas of the country, with the result that it has at times been arguing against experts from within its own ranks who promote a different causal interpretation of the role of speed in road crashes. Similarly, these alternative, contradictory voices have occasionally clashed with central government, disputing the evidence of causal processes put forward by proponents of the scheme, particularly in the early years of the NSCP (see, for example, BBC News Online 2003b, 2003c, 2004a, 2005c).

In one example from 2003, a vocal opponent, the then Chief Constable of Durham constabulary area, was criticised for refusing to install speed cameras in his area. The then Road Safety Minister drew attention to the ultimate measure of the accuracy of different causal interpretations when he noted:

> The chief of police would perhaps have to explain to local people why it is from 2001–2002 casualties in terms of deaths and serious injuries, and particularly those to children, it is actually one of the few areas where they are going up rather than down ... He will have to explain why deaths are rising in this area. (David Jamieson, quoted in BBC News Online 2003c)

This Chief Constable maintained, however, that speed cameras would not be an appropriate intervention. He noted:

> What we find is that there is not a single location within the county where you could say speed cameras would be useful in addressing a road casualty problem. We are still seeing the level of road casualties in County Durham are 33 per cent below the national average. I sympathise with every person who is a victim or family of a road accident in this county or anywhere. But I am trying to put my resources to the best use to reduce road casualties. (Paul Garvin, quoted in BBC News Online 2003c)

The core aim of preventing road death is used by both experts in support of opposing arguments. The Road Safety Minister was able to use statistics showing a measured increase in harm to support his view that cameras should be installed, while the Chief Constable of Durham was able to suggest that it is actually by *not* installing cameras that he best fulfilled his obligation to potential victims and their families. This is again achieved through the use of statistics that, in this case, promote the increase in casualties as 'a blip' (BBC News Online 2003c) and draw attention to his area's generally below-average statistics.

Further dissenting voices from within the police have also been critical of the use of speed cameras and have used the media in various ways to communicate their opposition and the reasons for it. The 'police' aspect of their identity is emphasised to lend credence and status to their maverick stance on the issue of speed cameras.

One senior officer, described in a Sunday feature as 'the man who brought speed cameras to Britain', was quoted as saying, '[E]ven I think they've gone too far', citing the potential for cameras to be used to generate revenue rather than increase safety as his justification for this claim (*Mail on Sunday* 2004). Another, billed as an 'insider' and former Partnership member by the *Sunday Telegraph*, was also given as the source of evidence that this financial motive underpinned enforcement (*Sunday Telegraph* 2004). Numerous other news stories attempt to gain credence through using this method, while the Association of British Drivers dedicates a section of its website to comments from the police that it takes to be anti-camera (ABD 2010).

Dissenting voices even appear to emanate from within the ranks of the status experts (and those who draw on their previous status within this category). Even those with the same mandate for control and the same goals in terms of casualty reduction (constrained by and accountable to the same targets) can come up with incompatible programmes for their attainment in the context of demonopolised expertise. This inconsistent expertise emanating from within the ranks of status experts demonstrates the extent of demonopolisation, and exemplifies the lack of an über-expert to whom the layperson can turn for guidance via which to 'negotiate the complex realities of modern life' (Roberts 2002, 255).

Representatives of Motorists – Implicated Instigators of Risk

In any risk debate the identification of a harm and an interpretation of its causation produces an implicated instigator of it (Beck 1992, 29). Given the context of demonopolised expertise, however, those so implicated may decline to accept the interpretations of the definers that result in their problematisation, and may promote their own alternative interpretations.

In this specific policy context, it is the driver and his/her car that is assigned responsibility for instigating the risk that leads to road death and injury. This section considers the organisations that have a significant presence within the debate through their claims to represent drivers as risk-producing agents and their vehicles as risk-producing technologies. It is via these representative organisations that the voice of implicated instigators is most commonly heard within the debate, and it is through these organisations that their individual views achieve expert validation and credence. Such organisations do not base their claim to expert status around their acceptance of the role as accused instigators, but claim expert status through their experience of the causal interpretation through which an increased risk of harm is alleged to result. Such representatives of implicated instigators frame their stance as *expertise*, allowing them to appear not as lobbyists but as experts whose interpretation happens to benefit their clients. Their case is put forward as resulting from 'consultation' as to public reception of the technology, or as 'research' into the effects of it on road casualty figures. Both these terms imply scientific exploration of the issue and are less amenable to the suggestion that they are simply biased against any restriction of the motorist. Their views are, or so it is claimed, backed up by science; not by bias in favour of any interpretation in particular, but by research.

The RAC Foundation and the AA Motoring Trust[13] were two motorists' organisations that featured prominently in the debate, representing motorists as official government consultees on all motoring-related policy, and offering responses to most emergent policy documents and strategies. It is this role as consultee that both organisations have promoted, above that of their role as lobbyists. This secured them a credibility derived from their (state-acknowledged) expertise, not from their role as interested parties with customers' interests to promote. This role has been established over almost a century, with these organisations (in their various forms) historically being invited to comment on policies that affect motorists (Plowden 1971). Both organisations claimed to represent the voice of the motorist through their large memberships, which were often surveyed for their opinions (see, for example, RAC Foundation 2004; AA Motoring Trust 2005). Both have also conducted their own research into motoring issues (see, for example, RAC Foundation 2002; AA Motoring Trust 2003) and frequently issue press releases on various motoring topics.

The expertise of these organisations has consequently been based on their representation of drivers and their familiarity with their views, but also on a stock of research evidence that supports their views and, in doing so, promotes the interests of their members. Both organisations have issued position statements setting out their stance on the issue of speed cameras as speed enforcement technologies, and these positions are considered below.

The RAC Foundation adopted a cautiously supportive approach to speed limit enforcement, raising issues of concern about the appropriate siting of cameras, and about traffic police 'concentrating on offences which can be easily detected by cameras such as speed, rather than the more serious offences which cannot' (RAC Foundation 2003). In this second sense, they have questioned whether this particular cause of risk deserves the priority it has been given and also the method of intervention that has been chosen by the authorities. The Foundation is often cited in press reports of camera-related issues (see, for example, *Daily Telegraph* 2010b) and commenting on concerns about the accuracy of cameras that are based on actual examples of faulty or improperly used equipment (see, for example, RAC

---

13   At times, the identities of these expert voices are confused by the existence of a variety of organisations affiliated to the AA or RAC. Where either is mentioned in media reports, by other experts, or even in official documents such as Select Committee reports (see TLGR Committee 2002, 79), it is not always possible to identify whether the body referred to is a Foundation, Trust or commercial arm of the organisation, all of which have adopted an individualised stance on the policy. For the purpose of the debate, it is the RAC Foundation and the AA Motoring Trust (until January 2007 when the IAM Motoring Trust assumed responsibility for its work) that are considered to represent the significant voices during the peak of the speed camera debate.

Foundation's Head of Traffic and Road Safety, in *Daily Mail* 2005). The main concern of the RAC Foundation has been, however, that the use of speed cameras is damaging relationships between the police and motorists. They have claimed that 'public perception of speed cameras is becoming increasingly negative', largely driven by the perception that 'speed cameras are more connected to raising revenue than reducing accidents' (RAC Foundation 2005). Research conducted by the RAC Foundation noted, for example, that three-quarters of drivers, when asked, said that they would not report a speed camera vandal to the police (RAC Foundation 2003). The Foundation has therefore voiced its criticisms of speed cameras through the words of the drivers it represents, citing its own research as evidence of the stance it puts forward.

At the peak of the debate, the AA Motoring Trust shared its rival's views on the placing of cameras, urging that they be placed only where excessive speed has caused accidents: 'The AA Trust supports the use of cameras primarily as a means of deterring speeding at *sites where the wrong speed can kill and seriously injure*' (AA Motoring Trust 2003, 3; emphasis in original). In stressing this caveat, it raised questions about the use of the technology in places where other causes were implicated in harm causation, and where it was improbable that speed cameras would have any positive effect on crashes. In doing so, it implied that some of the instances of detection, prosecution and punishment of its members will have been unwarranted, having taken place in locations where speed is not linked an increased risk of harm. The Trust also raised concerns about the impact of the policy on police/motorist relations, advising that '[c]ameras must not be deployed in a way that the enforcement of speed limits is perceived to be for revenue raising rather than casualty reduction. It is to be regretted that many drivers now perceive that cameras are there to raise money' (AA Motoring Trust 2002). This observation suggests an alternative motive for enforcement that is not based around its members' role in the causing of risk, challenging interpretations that justify the policy on the basis of a connection between speed and harm.

Despite its lack of heritage in comparison to motoring organisations like the AA and RAC, the Association of British Drivers (ABD) has also been a significant expert voice in this debate. The ABD effectively spans the gap between expert and audience, promoting itself as being run 'by drivers and for drivers', and boasting the slogan 'The Voice of the Driver'. The ABD gains a level of expert influence and status by virtue of having formed an organisation to represent the motorist on a variety of motoring issues. It also boasts 'spokesmen' [*sic*] who have a high level of training and qualifications *as motorists*. Their Road Safety Spokesman was, for example, described in a 2005 press release issued by the Association as being a 'Senior IAM[14] observer and RoSPA-qualified[15] driver and motorcyclist' (ABD 2005a). Such a description literally qualifies him to stand as an expert

---

14 Institute of Advanced Motorists.
15 Royal Society for the Prevention of Accidents.

representative of lesser-trained drivers who nonetheless, it is suggested, share the same concerns.

The Association's activities are increasingly dominated by anti-camera and anti-enforcement campaigning. It is officially an 'insider' (Baggott 1995, 19) pressure group, given that it is a member of the Parliamentary Advisory Council for Transport Safety,[16] but the usefulness of its contribution has been questioned by that organisation (*The Guardian* 2004a). Other commentators have described the Association as 'rent-a-quote' in relation to anti-camera campaigning (Chief Constable 1), or '"the provisional wing" of motoring organisations such as the RAC and AA' (*The Guardian* 2004a). With 21 references in the newspaper articles surveyed, the Association's presence in this area of debate rivalled that of the AA according to this measure, and it frequently appears as a critical voice in BBC online reporting. A strong web presence has also been secured by the ABD, with their website operating as a source of statistics and research evidence that contradicts the official line on the effectiveness of speed cameras.

Relationships between the ABD and some other expert groups were strained. This particularly applied to road safety organisations, as it was felt that 'Brake hate us, that's quite literally. I think RoadPeace do as well'[17] (interview with ABD road safety spokesman). Links with other more specifically anti-camera organisations exist, with the exchange of statistics and other research findings being commonplace.

The ABD's opposition to the use of speed cameras centres on its belief that the policy has been motivated both by a desire to discriminate against drivers and by a desire to raise revenue. It supports these claims by reanalysing and reinterpreting the government's own statistics and research on the role of speed in crash causation and the effectiveness of speed cameras (see, for example, ABD 1999a, 2002). It promotes alternative causal interpretations which suggest that the policy *causes* rather than prevents crashes by making drivers drive less safely, supported by data that it claims shows an increase in road deaths since the introduction of the policy (ABD 2003a).

The ABD also maintains that the findings of the two-year evaluation of the scheme were 'rigged'. It claims that the areas chosen to participate in the scheme had above-average fatality and casualty figures at the start of the trial and that they consequently could be expected to fall in subsequent years. It suggests that this 'regression to the mean effect', rather than speed cameras, accounts for the observed reductions noted in the report (ABD 2001). This allegation is made frequently and has become one of the stock criticisms used by the Association (see, for example, ABD 2003a, 2004a, 2004b).

---

16   A registered charity which advises MPs and Peers on road, rail and air safety issues.

17   Brake and RoadPeace are prominent road safety groups within the debate. Their role is explored below.

The ABD's dissection of the use of TRL 323, considered in more detail below, is also a frequently used criticism deployed to challenge the legitimacy of the scheme and to attempt to construct speed camera use as fraudulently motivated (see, for example, ABD 1999b, 2002).

In addition to the multi-issue motorists' representatives detailed above, single-issue pressure groups have been established with the sole purpose of campaigning and acting to undermine the NSCP and speed camera use more widely. Rather than focusing on the interests of motorists generally, they have a specific and directed objection to the enforcement of limits by camera, often believing that they are 'revenue-raising' tools for the authorities and maintaining that the causal interpretations on which the policy is justified are flawed. This is typified by the naming of one campaign 'Safe Speed'. Common elements of the websites of such organisations are the publicity of camera locations, advertisements and features offering radar and camera detection equipment, and methods of appealing against and avoiding speeding fines. The reporting of vandalism against cameras is also a popular function.

Speedcam.co.uk and safespeed.org.uk, the two web-based anti-camera organisations to be profiled here, maintained similar content to that of the ABD site, although it was specifically camera- and enforcement-related. Speedcam. co.uk publicised instances of speed camera vandalism, changes to speed limits, methods of avoiding detection, and the sites of other similarly minded organisations and individuals. It also began to reveal the sites of cameras some time before the Partnerships themselves chose to do this.[18] The site was described as 'dedicated to safety and safe driving' and displayed 'photos of the abuse of the "fair" speed camera partnerships'. The site also acts as host to the press releases of Motorists Against Detection (MAD), an organisation that engages in vandalism of speed cameras.

Until his death in 2007, the author of the Safe Speed campaign (www.safespeed. org.uk) came to be known as the anti-camera movement's statistician, featuring on televised interviews and in other media in the role of critic of government statistics (see, for example, BBC News Online 2005d; Monbiot 2005). His site acted as a repository for anti-camera research and analysis, with his expert credentials quantified in terms of the '5,000 hours' spent studying the research literature and the '350,000 words' that made up his site. The compliments of other commentators were also included, for example the testimony of 'the Editor of the Observer' and 'a Professor hired by Radio 4' (Smith 2006a). The aims of the site are described as 'to objectively analyse the evidence for various strategies for road safety improvement ... starting from a position where we mistrust speed limits as a primary road safety strategy' (ibid.). This site contains a large amount of reanalysis of the official case for speed camera use, under headings such as: 'The

---

18   The availability of speed camera locations, started by these organisations, has now become legitimised by the Partnerships who also offer this information and by the inclusion of fixed speed camera locations in road atlases and on satellite navigation devices.

case against speed cameras', 'The great speed camera con trick' and 'How road safety really works'. Its main criticisms of the official case were that speed limits could not dictate what is a safe speed in any set of circumstances, that both the statistics drawn from TRL research are flawed,[19] and that any apparent reductions in crash occurrence can be accounted for by the principle of regression to the mean noted earlier.

Injured Parties

Given that the identification of a harm is in some sense the catalyst for an ensuing risk debate, the role in the debate of those so harmed is considered significant. Although the choice of drivers and cars as the instigators of risk in this debate is hotly contested, there is general agreement that a real harm exists in the form of victims of road crashes. The fact of death or serious injury does, however, limit the effectiveness with which these victims can organise and represent themselves, and the situation is further complicated by the fact that instigators can, at times, become injured parties themselves. Although a neat division between victims and instigators is therefore not possible, the two *are* generally represented *as though* they are mutually exclusive categories.

Organisations established to campaign on various fronts in the name of road safety have a significant role in the debate. The most significant groups to emerge from this research were the organisations Brake, the Slower Speeds Initiative, RoadPeace and Transport2000,[20] all of whom offered general support for the government's policies. Such groups claimed their expert status by virtue of their familiarity with the harm that they perceived as resulting from the causal interpretations favoured by the government and to which they, on the whole, subscribed. These organisations are often used by the media to juxtapose the input of associations like the ABD or other specifically anti-camera groups (see BBC News Online 2003d; BBC Breakfast News 2003). Responses and statements are both statistical and research-based, often supplemented with an example of a real-life incident or specific victim (see, for example, BBC News Online 2003d; BBC Breakfast News 2003; BBC News Online 2005f). In this sense, their expertise has a grounded feel to it and gives names and faces to the abstract reasoning and statistical averages that make up the core of the debate's expertise.

Brake is a national road safety charity that offers both a campaigning and support function on behalf of victims. It fully supports the use of speed cameras, and opposed the requirements for crashes to have occurred before cameras could be installed, believing that cameras should be installed wherever a need was perceived (Brake 2004, 3). Brake supports the use of speed cameras to prevent both

---

19   The 'one-third' claim extrapolated from TRL 323, and the claim that a 1 mph reduction in speeds leads to a 5 per cent reduction in crashes drawn from TRL 421.

20   Now the Campaign for Better Transport.

pedestrian and driver deaths, so in a sense speaks for potential risk producers as well. Examples used in debates, however, tend to be pedestrian casualties and are often children whose role as victim is uncomplicated by their dual role as potential risk instigator (see, for example, BBC News Online 2003d; *BBC Breakfast News* 2003; BBC News Online 2007b). Brake has evidenced its position by research sourced from both academia (for example, Ashton and MacKay 1979; Stradling, Meadows and Beatty 2004[21]), the TRL (reports 323, 421 and 511) and government (including DETR 1999b; DETR 2000a; DfT 2003a; DfT 2003b) and defends the use of TRL 323, criticising 'anti-speed camera campaigners' for 'misquoting' the research (Brake 2004, 3).

The Slower Speeds Initiative (SSI) had the stated aim of highlighting 'the impacts of speed through sound research and effective advocacy', emphasising the evidential basis for its position (Slower Speeds Initiative n.d.). It was supportive of the government's policy and opposed attempts and moves to limit where the technology can be used. It referred to TRL and government research throughout its website, with the finding of TRL 421 (that a 1 mph reduction in speeds leads to a 5 per cent reduction in crashes) its favoured statistic. Jointly with the Parliamentary Advisory Council on Transport Safety (PACTS), it issued a deliberate rebuttal of challenges to the use of speed cameras, entitled *Speed Cameras: 10 Criticisms and Why They Are Flawed*. This publication engaged in a step-by-step deconstruction of the major criticisms of speed cameras, including allegations of revenue raising, reductions in traffic officer numbers, illogical siting of cameras and the suggestion that cameras cost, rather than save, lives (PACTS and SSI 2003).

Along with the SSI, the campaign group Transport2000 made a legal challenge to the government's proposals to make all cameras more visible, on the grounds that this would reduce their effectiveness (Transport2000 2002). The organisation claimed that:

> Lobbying from motoring organizations has, however, led to a Government climbdown over where cameras can be sited. Current guidelines state that cameras should be highly visible to motorists and can only be installed at places where there have been four or more deaths or serious injuries or a heavy crash history. This has led to concern among road safety groups that motorists will simply slow down when they see a camera and then speed up again afterwards and that communities living on roads with a speed problem may be denied cameras because not enough people have been killed or injured. (Transport2000 2003).

A similar concern was voiced by the national charity RoadPeace, which, in its submission to the Transport Committee on Traffic Law and its Enforcement, noted that speed cameras were subject to a disproportionate amount of restrictive legislation: 'There are greater restrictions on safety cameras than there are on

---

21   Referenced as 2002 in Brake (2004).

CCTV, i.e. greater priority is given to protecting property than to preventing death and injury. It is permissible for plain-clothed detectives to be used to deter thieves, but safety cameras must be highly visible and signed in advance' (RoadPeace 2003).

Despite these reservations about the actual deployment of cameras, RoadPeace also fully subscribed to the causal interpretations favoured by camera proponents, offering links to official research and issuing press releases celebrating the reductions in casualties evidenced in the NSCP evaluation reports. In this sense, road safety organisations generally object to what they see as the government's vulnerability to pressure from the motoring lobby, while the motoring lobby would argue that the opposite is the case and that road safety campaigners have obtained an undue influence over the government. This illustrates the competitive nature of the risk debate, with instigators and injured parties competing to dictate what action is taken with regard to intervention and control, and in which direction it is aimed.

Academic and Other Research Evidence

The notion of scientific expertise as a vital witness on behalf of policy-making and policy-enforcing agencies has created a significant role for its authors. This role has been made clear in the above analysis where research is sought, cited in and commissioned for inclusion in policy documents. In the case of the Labour government 1997–2010, this related to research that supported causal interpretations which showed a connection between exceeding speed limits and an increased likelihood of crash occurrence, but such scientific expertise has also been used as evidence for opposition to such a relationship. Such scientific expertise is therefore called upon and produced to evidence contradictory interpretations, and reinterpreted subsequently to challenge or undermine an initial interpretation.

Academic research input of this type has been drawn from a variety of disciplines including transport psychology (see Stradling), sociology (see Buckingham), transport planning (see Keenan), law (see Corbett), and engineering (see Leeming and Mountain). It has also been co-opted into this current debate from other time periods[22] and other, international, settings.[23] Scientific expert and academic input has been used to claim to prove (Taylor et al. 2000) or disprove (Corbett 1995, 345; Davis 2002) the relationship between higher speeds and crashes, and also to prove (Gains et al. 2004), disprove (Buckingham 2003) and cast doubt upon (Keenan 2002, 2004) the effectiveness of cameras at reducing speeds and/or crash

---

22   For instance, J.J. Leeming's 1969 book *Road Accidents: Prevent or Punish*, used by the ABD (2005b), and frequent references to the impact of the 50 mph limit introduced following the Suez crisis (see, for example, Plowden and Hillman 1996, 32).

23   For example, Australia (Buckingham 2003) and the United States (Corbett 1995, 345).

incidence. For the purposes of this research, the reasons for the differing findings are not as important as the fact that they exist and are deployed by the various sides within the debate as scientific evidence of the case for which they wish to secure support. The methodological or other reasons that may account for the differences in findings are not, it is suggested, likely to be grasped by ordinary members of the public audience for the debate, but as a whole they are witnessed by those seeking out expert guidance on the issue of speed limit enforcement. As the *Daily Telegraph*'s 'road safety expert' noted somewhat exasperatedly: 'The trouble is, depending on where you sit in the debate, you can turn virtually any of the evaluations, studies and academic reviews of existing reports into a sparkling endorsement of speed cameras, or a total damnation' (*Telegraph Motoring* 2005, 5).

The following review of relevant speeding-related research indicates the ways in which it has contributed to the debate, the occasions on which it has been used by other debating experts, and the potential areas for critique that remain. First, however, the issue of Transport Research Laboratory Report 323 is considered in more detail. It is presented as a case study of the use of scientific research findings by those in favour of and those opposed to speed cameras.

TRL 323

One of the most important pieces of research with a particular role in the early years of the debate was, as has been suggested above, the report by the Transport Research Laboratory entitled *A New System for Recording Contributory Factors in Road Accidents* (Broughton et al. 1998). This report, known within the debate as TRL 323, was the subject of considerable discussion and figured prominently in both the construction of the official case in favour of speed camera use and the opposition's attempts to deconstruct that same case. Although no longer used in official justifications of the NSCP, or to support the subsequent use of cameras, the study was central to the initial case for prioritising speeds and for using speed cameras, and continues to form a significant element of criticisms of that case. The significance of the study as a focus for analysis by both sides of the debate means that some acknowledgement of the criticisms and defences of it is necessary and instructive.

Confusion about the meaning and significance of TRL 323 occupied those involved in the debate at its outset and continues to surface years after the introduction of speed cameras to the UK's roads (see Booker and North 2007). The disagreement centres around whether speed is implicated in around 7 per cent or around 33 per cent of all crashes, and thus whether or not it is legitimate to claim that speed cameras are likely to contribute to sizeable reductions in road crashes. The confusion arises due to disagreements as to the validity of adding other reasons for crashes detected within the study to those associated with speed. Factors such as 'aggressive driving' and 'driving too closely' have been added to 'pure' excessive speed (of the type that speed cameras can detect) to produce an overall figure of accidents in which speed may be implicated. Opponents have argued that such additional factors have been added inappropriately (ABD 2002), while supporters point out that speed exacerbates the consequences of all these other factors (Slower Speeds Initiative n.d.; Taylor 2002, 3).

Although TRL 323 was used extensively in early policy-making and other official publications associated with the NSCP, it was subsequently dropped in favour of later research, for example TRL 421. This later research was given prominence in the *New Directions* review (DETR 2000b) and the two-year evaluation of the NSCP (DfT 2003a), and is the preferred TRL option in more recent discussions (for example, DfT 2006a, 2006b). This change in emphasis and evidence base was noted with some alacrity by the Association of British Drivers, who described it as a 'climbdown' in a press release referring to its replacement headed 'Foundation stone of "Speed Kills" abandoned by DfT'. In this press release, a DfT spokeswoman was quoted as saying:

> The Department has, in the past, suggested that around one-third of accidents are speed related. This is not a figure it continues to use. But not because the Department no longer believes in its accuracy. Just as speed is a complex issue, so is the recording of contributory factors. (DfT Spokeswoman, quoted at ABD 2004c)

The Association, revelling in the change of emphasis, then notes that '[t]he words may be carefully chosen but the message is clear – the evidence from real accidents don't support the "one-third" fallacy, and the DfT have been forced to abandon it' (ibid.). The Association of British Drivers produced a dissection of the original study in 2002, but continues to attack the official use of it on its website. Its original critique notes that:

> TRL 323 showed that excess speed was a factor (but not necessarily the only factor) in **7.3 per cent** of accidents. Of course, this small problem did not divert the DTLR, who were by this time politically welded to the great 'one-third' lie. They got around their little problem by defining all sorts of other accident causes as 'speed related' until they managed to get them to add up to a third. (ABD 2002; bold in original)

Further criticism has come from the Safe Speed campaign website, where its author devotes considerable space to challenging official interpretations of TRL 323, including numerous emails to the DfT and TRL in which figures and claims are challenged. One document, 'Speed cameras – The case against', notes:

> For almost a decade the Government has been claiming that 'one third of accidents are caused by speed'. This absurd claim has no foundation in scientific fact, although the TRL have disgraced themselves by attempting to justify it

in print. The truth is that a very small percentage of accidents are caused or contributed to by speed in excess of a speed limit. (Smith 2004b, 5)

Not unaware of the criticism being levelled at TRL 323, the first sentence of the government's response to the Transport, Local Government and the Regions Committee's Road Traffic Speed report (Secretary of State for Transport 2002) is devoted to defending its evidence base for the importance of speed in crash causation:

> Those critical of speed management often misquote or selectively quote from TRL Report 323 … to argue that it is wrong to claim that speed is a major crash and injury causation factor. However, this TRL report has been persistently misquoted and used out of context. If the report is read in its entirety, it clearly shows that the factors that comprise driving at both excessive and inappropriate speed effectively confirm the one-third figure of speed being a contributory factor in road accidents … *The proof that excessive speed is a major cause of crashes may be further found when speed management measures have been put in place and significant reductions in fatal and serious casualties occur.* (Ibid., 3; emphasis added)

In this way, the success of the NSCP is itself used retrospectively to prove that the rationale that justified it in the first place was correct.

Such was the importance of the controversy over TRL 323 that, in 2002, the TRL published an article under the headline 'Speed and accidents – let's put the record straight!'. It noted:

> There is a vast amount of evidence demonstrating the strong link between vehicle speed and road accidents. So why does material keep appearing in the media suggesting the effect is small? The issue is so important we feel it is time to reiterate the true position. In the 1990s a number of police forces conducted a limited trial of an experimental accident reporting system. The results were reported clearly in TRL Report 323 but they have frequently been misquoted … Misunderstandings in the press appear to have resulted in two ways. First, speed identified as a separate factor in its own right was present in 15 per cent of accidents, not the 7.3 per cent, or lower figures, that are often wrongly quoted. Secondly, the 15 per cent is only one part of the total effect of speed on accidents. When allowance is made for all the other speed-dependent factors, the contribution is, we believe, much greater. (Transport Research Laboratory 2002, 3)

The clarifying article also points out that other TRL studies have explored the relationship between speed and crashes since the controversial study: 'These studies together provide extremely robust evidence of how speed affects accidents. They are large-scale studies, of real traffic on real roads involving rigorous statistical analysis. The results are unambiguous. Remember, 10 people die and 100 are seriously injured on our roads *per day*' (ibid., 3; emphasis in original). The reference to the core aim of all those interested in road safety – saving lives – reinforces the idea that reducing speeds and preventing death are synonymous. Questioning the accuracy of *this* research is therefore presented as denying the importance of preventing road death and injury *generally*. The counterclaim, that tackling speed is not the most effective or efficient method of achieving this agreed aim, is thus obscured. The article also makes clear the evidential basis of its claims, based on 'real' circumstances and maintaining, despite the fervent debate that necessitated its production, that the results are unambiguous.

Academic Authors and the Speed Debate

The two authors whose work is considered first, and in most detail, were identified as significant in several ways. The work both of Claire Corbett and of Steve Stradling is, first, referenced with some frequency in policy documents, committee reports and other official publications. Second, it has the most noticeable history of securing a media response or some comment from elsewhere in the debate when it is published. Third, both authors were named by a senior DfT official during an interview for this research as being two of the 'names' that would be 'invited in to the DfT' for a 'closed-door seminar' on issues relating to speed camera use (interview with senior DfT official 2004). The consideration given here to their work of direct relevance to the speed camera debate is not intended to be a detailed deconstruction of the methodologies and findings, and I deliberately stop short of giving my own views on the research. The reasons for this approach are set out in the introduction to this work. Rather, this section considers the most publicly accessible aspects of their work and how their work has been used by different parties in the debate to different ends.

Corbett and Simon's study, 'Police and public perceptions of the seriousness of traffic offences' (1991), involved a sample of the general public and a sample of police traffic officers rating the seriousness of various traffic offences, including various speeding offences. The study predated the NSCP, being published in the same year as speed cameras were first approved, and as such accessed public and police opinion prior to the controversial expansion of the scheme under the NSCP. It was cited as evidence in DfT policy formation in 2002, with specific attention paid to the issue of speed limits and their transgression (DfT 2002). In reviewing literature from the 20 years before, Corbett and Simon had found that general support for speed limits was high, as well as for the view that the roads would be safer if drivers kept to limits (1991, 159). They also noted, however, that other authors found that drivers were generally more critical of other drivers speeding than they were of their own speeding behaviour (ibid., 159). The most seriously regarded speeding offence was found to be that of driving at over 50 mph in a 30 mph zone. However, driving at between 71 and 80 mph on a motorway was both the least seriously regarded speeding offence and the least seriously regarded traffic offence overall (ibid., 157). The study noted that, in 'taking a dim view on exceeding 50 mph in a 30 mph area, the police are consistent with public opinion' and that consequently attempts to enforce against such offenders would potentially be well received. However, the offence of exceeding a 30 mph limit by between 31 and 40 mph was rated one of the *least* serious of all the 26 offences by both public and police. It is just such offences that would subsequently become the focus of the NSCP, however.

In a study published in 1995, Corbett surveyed the attitudinal and behavioural effects of one of the earliest trials of speed cameras on drivers. This self-report study, which formed part of the West London research (London Accident Analysis Unit 1997/2003), found that some types of driver were more inclined to modify their speeds than others, that speeders showed more general awareness of camera locations than non-speeders[24] and that fast drivers wanted fewer cameras while slower ones wanted more (Corbett 1995, 350). It also noted reductions in accidents, although the parallel observation that rates also fell on roads without speed cameras is also made. This is explained by the suggestion that the cameras brought about reductions in speeds in the general area, not just in the immediate vicinity of the camera, although the study suggests that only 4 per cent of drivers claimed to be changing their behaviour in this way (ibid., 350).

Later research by Corbett and Simon continued this interest in the responses of drivers to automated enforcement (Corbett and Simon 1999). The research involved nearly 7,000 drivers who took part via a postal survey. In-depth interviews were conducted with a sub-sample of 105.[25] The study set out to explore the 'effects and effectiveness of various strategies related to the deployment of speed cameras' (ibid., 2) and was conducted between 1993 and 1996. It proposed a typology of four types of drivers who each responded differently to speed enforcement by speed camera. 'Conformers' were unaffected by enforcement as they already obeyed speed limits, the 'deterred' reduced their speeds on roads involved in the study, 'manipulators' slowed down for cameras and sped up afterwards, and 'defiers' continued to drive at above the speed limit (ibid., 10). The last two categories, focused on by the research, contained those with the highest crash rate and those who tended to deny an association between speed and crash risk (ibid., 19). The report concluded that, despite the variety of responses to speed cameras, to some degree 'everyone has a price' and that 'sooner rather than later' the numbers of drivers who failed to be controlled by them would reduce (ibid., 70). It also recommended that 'future media campaigns should point out that most drivers think they are better and safer than others which is illogical, and that the message of the danger of speed is directed at all drivers and does not exclude those who believe they are better' (ibid., 73).

---

24   There was no requirement for cameras to be highly visible at this time.

25   For methodological data relating to the first survey, see Corbett and Simon 1999, 12–17.

The most referenced aspect of the study, within the debate, has been that which set out to 'assess the strength of drivers' positive and negative views towards speed cameras' (Corbett and Simon 1999, 53). The survey asked 6,879 respondents to agree or disagree with eight statements. The four given below have subsequently been reused in all three NSCP evaluations (see DfT 2003a; Gains et al. 2004, 2005):

'Cameras are meant to encourage drivers to stick to the limits, not punish them.'

'Fewer collisions are likely to happen on roads where cameras are installed.'

'Cameras are an easy way of making money out of motorists.'

'Cameras mean that dangerous drivers are more likely to get caught.'

Responses to these questions have been referred to elsewhere, and a consideration of the changes found over time can also be found in the three evaluations of the NSCP (see DfT 2003a, 5-1–5-3; Gains et al. 2004, 44–50; and Gains et al. 2005, 62 and 73).

Various elements of the work of Corbett (et al.) have therefore contributed to the official evidence base in support of speed cameras. Other elements have, however, been cited as evidence 'that the causal link between speed and accidents is unreliable at best' by the ABD in their response to government consultation (ABD 1999c, referring to Corbett and Simon 1992). The initial research by Corbett into the impact of the West London project (London Accident Analysis Unit, 1997/2003) is also noted by another academic in his construction of a case against the use of speed cameras. The caveat that 'present "contentment" could evaporate, to be replaced by alienation of the average driver' (Corbett 1995, 353) is highlighted as providing evidence of the potential for automated enforcement to erode the 'goodwill' on which the police rely (Buckingham 2003, 10). Corbett and Caramlau's 2006 research into gender differences in responses to speed cameras was criticised by Safe Speed for conflating the issue of road safety with that of speed cameras. As such, the criticism is less of the findings of the report than of the underlying assumption that speed cameras are a road safety intervention, and the objection seemingly only uses the publication of the report as an opportunity to repeat the usual concerns of the author (*Daily Mail* 2006a). Various aspects of this academic's work have therefore been used by both sides of the debate depending on which particular case they are felt to support.

Professor Steve Stradling, at Napier University, previously of the Transport Research Institute, was another academic named by the senior DfT official interviewed as part of this research as being invited to discuss speed camera-related policy at a closed meeting in 2004. His work was cited extensively in proceedings from DfT seminars on behavioural research in road safety (see DfT 2003c, 2004a, 2005c, 2006c, 2007) as well as in other official publications. Several of his studies feature on the DfT website, and he made numerous media appearances during the lifetime of the NSCP (see, for example, BBC News Online 2004b, 2006b). His work of specific relevance to the NSCP is therefore briefly reviewed here.

Stradling is responsible for the concept of the 'crash magnet' driver. Such a driver is, it is proposed, particularly at risk of being involved in a car crash because of a proclivity to violate safe driving practices, including breaking the speed limit (Stradling 1997). It is *violations* specifically, rather than errors or lapses in concentration, that are linked to this increased risk (Parker and Stradling 2001, 8). The formula 'Violation + Speed = Crash' – a development of this research – was adopted by the 2000 DETR review of policy in relation to speed management as part of the 'overwhelming evidence' in favour of speed control (DETR 2000b, 8–9). It also featured in the Transport, Local Government and the Regions Committee on Road Traffic Speed (TLGR Committee 2002). The focus of the NSCP on the speed limit specifically is thus justified by this research, given that the limit creates such violations and represents the demarcation of 'safe' as opposed to 'dangerous' driving speeds. Campbell and Stradling's later DfT-commissioned research *Factors Influencing Drivers' Speed Choices* (Campbell and Stradling 2003) was described as a 'large scale study of Scottish drivers' and compared 'the responses to [various driving situations] of drivers identified as speeders and non-speeders, and the responses of those who had and who had not been involved in an RTA [road traffic accident] in the past three years' (ibid., 2). The study classified 'speeders' as those who had been detected by camera or stopped by a police officer in the previous three years. Subsequent analysis adopted the dichotomy of 'speeder' and 'non-speeder' to classify participants in the research. As well as being more likely to be involved in crashes, the report was able to conclude that 'speeders' were more likely to report that they slowed down for speed cameras, but were also more likely to speed when late or when traffic around them was speeding (ibid., 12).

The study has subsequently been referenced in the proceedings of DfT behavioural research seminars (DfT 2004a) where the connections made between having been detected speeding and crash involvement were noted. As such, the aspect of the research most noted is, again, that which supports an emphasis on the speed limit as the differentiation between 'safe' and 'dangerous' behaviour.

Stradling has also been involved in the production of a series of studies of 'the causes and consequences of speeding' for the Scottish Executive (Stradling et al. 2003). This research is also cited within the debate as evidence that challenges the allegation that cameras do not catch dangerous drivers (PACTS and SSI 2003, 4), and features in the Select Committee report *Traffic Law and its Enforcement* which subsequently made recommendations in support of speed camera use (House of Commons Select Committee 2004).

Other research by Stradling, published in 2006,[26] was condemned by the ABD as 'simplistic' and 'contrived', and its author denounced as 'an anti-car academic' (ABD 2006a). The same research was also received critically by the Safe Speed organisation, with their author noting: 'I believe that this research is nonsense, strongly influenced by pre-conceptions and vested interests. Speed cameras do not identify risky drivers nor do they make our roads safer' (Smith 2006b). Stradling's work appraising a local Speed Awareness Course[27] (2003) was, however, given a more positive reception by the same author, even being filed under the banner 'Talking Sense on Speed' (Smith 2003a).

Some research by academic authors has taken a view that directly and deliberately opposes that of the official research and much of that considered above. Although not a specific research project in itself, the sociologist Alan Buckingham's article 'Speed traps: Saving lives or raising revenue?' (2003) is a useful summary of the case against speed cameras and has been credited with giving 'academic credence' to and 'validating' the opposition arguments of non-academic contributors (Smith 2003b). It has also, however, been the source of criticism from pro-camera campaigners and academics, as described below. Among other criticisms, the article condemns the use of TRL 323, rehearsing the familiar criticism that the 'one-third' figure has been arrived at through the inappropriate addition of other crash-causing factors (Buckingham 2003, 4). Research that implicates *slow* drivers in risk causation is also cited (ibid., 6), along with the claim that British speed limit enforcement policies have resulted in the reversal of a trend that saw a year-on-year reduction in accidents of 3 per cent (ibid., 7). The latter claim is clearly presented as being the result of the NSCP:

> If we correlate the increasing 'fatality gap'[28] caused by the divergence between the 1966–1993 and 1993–2001 [road fatalities] trend lines with the rise in speeding convictions by speed cameras since 1993 we obtain an almost perfect correlation of +.97. In other words, there is an almost perfect linear relationship between the increase in speed camera tickets and the increase in the fatality gap.
> (Ibid., 7)

The article also raises concerns about the impact of speed limit enforcement policies on relationships between the police and motorists, predicting not only that 7.2 million motorists will be convicted of speeding by 2010, but the dire consequences of this:

---

26   A survey of drivers' attitudes conducted for the West Midlands Casualty Reduction Partnership (Stradling 2006).

27   An alternative disposal for drivers detected speeding by speed camera.

28   The 'fatality gap' is the name given here to the difference between the predicted and actual numbers of deaths resulting, per year, from road crashes.

> The danger is that motorists will notice that the mass conviction of speeders is being matched by a retreat from catching criminals. This risks alienating those on whose goodwill the police often rely. By regularly convicting large numbers of law-abiding people, it is also possible that respect for the law will lessen in other areas. (Ibid., 10)

The report concludes that speed cameras may be implicated in causing, rather than preventing, crashes, that they result in a 'de-skilling' of drivers who become less capable of accurately judging circumstances, and that reductions in road casualties have actually been the result of developments in engineering and car design (ibid., 11). As alternatives to speed camera enforcement, the report finally advocates discretionary policing of speed, the more stringent enforcement of laws against 'dangerous driving' and the 'responsibilization' of drivers in relation to their own contextualised risk-avoidance strategies (ibid., 11).

The Australian researcher Max Cameron subsequently challenged Buckingham's approach and the two academics then engaged in a debate hosted by *Policy* magazine. Their exchange – typical of the way in which speed limit enforcement has been discussed within the debate – is considered below in some detail. The resulting article was published under the title 'Speed off' (Cameron and Buckingham 2003). Under the subtitle 'Speed cameras work', Cameron criticised Buckingham's research, claiming that '[i]t include[d] much superficial analysis purporting to assess the effects of speed cameras in Britain and Australian States' (Cameron, in Cameron and Buckingham 2003). Having dissected and criticised each aspect of Buckingham's analysis, Cameron concludes:

> Subsequent sections of the article rely on Dr Buckingham's suppositions that speeding is a relatively unimportant problem, and that speed cameras are ineffective and even counterproductive. None of these suppositions is true, and MUARC[29] has provided evidence to the contrary. Hence no further comment on Dr Buckingham's opinions is necessary. (Ibid.)

In his defence, Buckingham then responded, under the subtitle 'Speed cameras not the answer', by again using the findings of TRL 323 to support the claim that excessive speed makes only a minor contribution to road death and injury, and noting the potential damage to police/motorist relations: 'Perhaps the most serious weakness of Professor Cameron's reply is his failure to address the unintended consequences of speed cameras ... Professor Cameron has failed to address the growing public opposition to speed cameras and the consequences for policing' (Buckingham, in Cameron and Buckingham 2003).

As such, it is clear that at all levels the creation of a sufficient scientific evidence base involves both research into the effects of cameras on casualty rates *and* their effects on public attitudes. A theme of this analysis has therefore been that

---

29   The Monash University Accident Research Centre.

'attitude' data has been produced and sought out alongside 'effectiveness' data. Such evidence is vital to a programme that ultimately relies on the compliance of the general motoring public to be effective, and is an essential aspect of the legitimacy of the programme. Damage to relationships between public and authority is therefore considered to be a significant 'risk' in itself. A new potential harm, and a new proposed set of causal interpretations of the processes from which it results, has therefore emerged over the course of the debate. It is now necessary that authorities try to prove that the technological intervention they have chosen is not itself a source of manufactured risk.

Conclusion

The risk context in which the debate takes place can be seen to have created both an increased reliance on scientific evidence and an increasingly important role for experts who claim to be able to interpret and apply it. The role of experts is, however, also profoundly affected by the risk context in which they work, to the extent that expertise has become demonopolised across a wide range of sources.

The various experts engaged in debating the rights and wrongs of speed enforcement demonstrate both the importance of scientific analysis in the preparation of their case and their awareness of the debate context in which they are engaged. Demonstrating that interpretations are both *accurate* and *accepted* as such then becomes the dual role of the expert engaged in attempting to secure definitive expert status in the speed limit enforcement debate.

The promotion of one expert interpretation above another can no longer, however, be achieved simply through the deployment of science, given that a variety of competing expert voices are deploying the same criteria and reaching contradictory conclusions, all with equal conviction. The increasing need for evidence, despite its inability to offer conclusive proof, is just one consequence of this debate among experts, however. Further consequences are considered in the next chapter.

Chapter 4
The Expert Marketplace

Having considered the way in which the various debating experts use (and, their opponents might say, abuse) scientific evidence in support of their position, this chapter proposes that science alone has proved insufficient in the construction of a persuasive expert case. It considers the resulting need for expertise to be *promoted* in order to stand a realistic prospect of being accepted by the public as 'the truth' in relation to this particular risk issue.

Second, the chapter considers the consequences of this marketplace full of expert interpretations for the 'consumer' – the member of the public faced with contradictory expert products. It explores evidence that suggests that members of the public have obtained a degree of expertise for themselves, gleaned not from hard science but from *experience*, that allows them to adopt their own views about speed limit enforcement in the face of the incompatible claims with which they are faced.

The Promotion of Expertise

Underlying the existence of the role of 'expert' are assumptions about the nature and characteristics of the audience for expert products. This audience is, Beck suggests, thought to be '*incompetent* in matters of their own affliction' (Beck 1992, 53) and ready and willing to be enlightened by the greater knowledge of their expert superiors:

> [The] division of the world between experts and non-experts also contains an image of the public sphere. The 'irrationality' of the 'deviating' public risk 'perception' lies in the fact that, in the eyes of the technological elite, the majority of the public still behaves like engineering students in their first semester. They are ignorant, of course, but well intentioned; hard working, but without a clue. (Ibid., 57)

The following quotations demonstrate the degree to which an assumption of ignorance on the part of the public underpins expert belief about the debate. The public is viewed as being essentially *persuadable* in relation to risk and its causes, given the belief that the public is dependent on experts for these interpretations. The quotations below relate to a planned Department for Transport project for the more effective dissemination of research findings, and a road safety group's recommendation for the same:

There appears to be a gap between the actual level of risk from inappropriate and excessive speed and driver perception of the risk. This project will aim to facilitate the development of a more effective strategy for informing the public about the risks of speed, and the basis for speed management and enforcement policy, in order to modify driver behaviour in relation to choice of speed and attitudes to enforcement of speed limits. This will include better means of dissemination to the public of existing research and knowledge about the role of speed in the cause and severity of accidents. (DfT 2006d, 7)

Police should be required to record the speed of travel of vehicles in fatal and serious injury collisions, so the contribution of speed to serious crashes can be identified and reported to the public. (Brake 2004, 4)

Clearly, information (and plenty of it) is viewed as the key to the 'modification' of public behaviour. That this 'supply' approach to convincing the public has not worked so far is a source of much frustration to those charged with winning the debate at practitioner level. In the following example, a Casualty Reduction Partnership member alludes to the official evidence base in favour of the use of speed cameras, but overlooks the existence of the competing interpretations (explored in the previous chapter) which are also available for the public's consideration:

It's frustrating because we have all the evidence and we don't hide it, but people still don't seem to want to believe it! I don't know why because it's there for them to see, but they still come back to us with this stuff about it being about making money! (Interview with county council Partnership representative)

The implication is that members of the public are expected to absorb the debate without prejudices, preconceptions or biases of their own, and will be susceptible to the publication of masses of data that purport to conclusively evidence a particular interpretation. This assumption relies on the existence of a 'passive expert–client relationship' (Lea 2002, 125), an assumption that underpins the production of the various publications considered in Chapter 3. This chapter considers how this assumption is reflected in the various strategies of promotion used by experts and other, wider, consequences of operating from this starting point.

Given the reliance on scientific proof within debates about risk, it has been suggested that other alternative discourses have been dismissed from the debate as 'merely anecdotal, uncontrolled non-knowledge' (Lash and Wynne 1992, 5). However, the demonopolised context can be seen to have created a situation in which competing scientific interpretations actually require seemingly quite unscientific alternative discourses to set them apart from each other. Given that all interpretations claim to be supported by pre-existing facts about the way the world works, science becomes '*necessary* for the production of knowledge, but [is] no longer *sufficient*' (Beck 1992, 169; emphasis added). Science needs, additionally, to be *marketed*: '[I]t is only the "extra" of presentation, personal persuasive power, contact, access to the media or the like which will provide the "individual finding" with the social attribute of "knowledge"' (ibid., 169). Therefore, although this debate is fought, at one level, via the exchange of scientific findings, at a second level a range of 'extras' are required to distinguish between conflicting scientific interpretations. This is the case because those who would seek to regulate the behaviour of others must effectively persuade the public, in this case, that they should consent to having their behaviour restricted for the sake of others. Those who disagree with this type of intervention must, on the other hand, persuade the public that this kind of restriction is not required. In this specific risk context, that audience is not only largely comprised of potentially implicated instigators but also broadly coterminous with the electorate who can ultimately vote for the causal interpretations that they find most convincing and against those that fail to be sufficiently persuasive. The assumed ignorance of the audience as to the scientific merit of each competing interpretation means that they cannot be relied upon simply to identify the best (meaning the most scientifically rigorous) interpretation and subscribe to it. The *promotion* of expertise is therefore necessary.

The following sections explore the ways in which Beck's 'extra[s] of presentation' (Beck 1992, 169) feature in this specific debate, by considering five strategies of persuasion and promotion which supplement the scientific justifications that all 'sides' in the debate claim to possess. These supplementary strategies consider the importance of access to the media, the use of experts *by* experts and the undermining of opponents' scientific capabilities. Two further strategies using 'homogenising' and 'marginalising' discourses to influence the audience and to exploit its vulnerabilities and commonalities are also explored. These persuasive strategies are all generally publicly unacknowledged tactics, with the pretence that the debate is simply an exchange of facts maintained by the experts quoted below. As such, the use of these tactics is partly evidenced here through the use of interviews with these experts which access the back-stage performances of the debate. The analysis conducted below therefore challenges the promotion of risk regulation as an objective, scientific or inevitable process of research, analysis and intervention, portraying it instead as the result of a conscious and determined effort to present it as such through the use of various strategies of persuasion. Attacks on it and the interpretations on which it is based can, similarly, be viewed as strategic and premeditated attempts to discredit it.

Strategy 1: Access to the Media

The need to reach members of the audience for the debate so that a particular interpretation can be put forward as definitive to them means that access to the media is vital. Beck identifies the media as essential to the effective transferring of a research finding into the public domain so that it can begin to shape reality:

'Access to the media becomes crucial ... Good arguments, or at least arguments capable of convincing the public, become a condition of business success' (Beck 1992, 32). Furthermore, Haggerty notes the way in which this privileged position that the media has secured results in changes to the debate itself. He notes that we have 'a media establishment that now expects that a social problem's dimensions will be quantified' (Haggerty 2002, 95), but which also has a notoriously short attention span when it comes to much of the scientific detail that underpins this ability to quantify problems. As such, facts and figures are necessary to secure an audience, with a percentage or an extreme figure a good way of attracting an audience's attention – hence the appeal of the poll or survey finding. 'Findings' are, however, stripped of much of their nuanced context (and potentially *meaning*) in order to be rendered palatable (and, ironically, *meaningful*) to a non-expert public. The use of sound bites, statistics and sensational data can be seen as a necessary strategy for ensuring that the story is headline-grabbing and can compete with the media output and representation secured by an opponent.

A second way in which the media exerts an influence over the debate is, however, more overt and deliberate. In the following example, a traditional 'definer' of risk expresses frustration at the role the media has performed in undermining his position as a proponent of the Safety Camera scheme:

> It seems astonishing to me that it could ever be a matter for debate, but it has become a matter of debate and there are interesting lessons to be learnt about how we got into this situation. We allowed a small but vocal group of people, who are either in denial of the evidence or aren't interested in the evidence, to take over the debate and persuade a significant section of the press – not just the tabloid press, I'd include some of the broadsheets in this as well – and a significant chunk of the population, that safety cameras are there to, quote, 'raise money', unquote. (Interview with Chief Constable 1)

The inability to secure control over the presentational methods favoured by the media was a source of frustration to the traditionally definitive central government and police experts in particular. They felt that opponents had managed to achieve an unwarranted level of publicity for their views simply because the media favoured a presentational form that encouraged debate concerning issues about which they felt they had the right to be the sole expert voice. If the media does not rely on scientific merit to determine its stance on an issue (but on such things as securing an audience), then the experts who claim to base their position on science alone will be frustrated by its portrayal of them:

> The ironic and irritating issue, though, is how much airtime they've managed to secure. Some of this is built into, for instance, the BBC's charter. I've had some very strange debates with producers in the BBC because they feel that their charter requires them to show balance. So they get rent-a-quote in from the ABD in order to transmit a balanced programme, but I then say, 'Well, hang on,

99 per cent of the population don't agree with this person.' It's always balance, balance, balance ... They are also seeking a thrill, of course; they want the sensationalism. It's not a very good debate on Radio 4, let alone Radio 5, to have everyone having a love-in and agreeing how wonderful cameras are. Because it doesn't sell airtime. (Interview with Chief Constable 1)

The demonopolisation of expertise means that the vociferousness and volume of a particular voice can be a determining factor in its access to the media, rather than the soundness of its evidence, the popularity of its stance or its previous reputation as an expert. The media, like the public, are unlikely to be in possession of sufficient skills to be able to assess the 'true' scientific merit of research findings brought to their attention. Furthermore, they are in any case likely to have their own agendas and concerns that determine which research they promote as valid and which they dismiss, as one expert was well aware:

The skewed debate run deliberately by the tabloid press is not even in denial of the evidence, but *in spite of* the evidence. I am well aware that the tabloid press is not managed by stupid people. They know perfectly well what the evidence says. They are not deluding themselves; they are simply following a line that they choose to follow, *despite* the evidence. I think they have successfully persuaded a very large number of people that this is in effect some sort of tax, and I think they are deliberately pursuing that line for political reasons. (Interview with Chief Constable 1)

Attempts to supply conclusive evidence are thwarted by the media's apparent disinterest in 'the facts'. Of course, the media is itself an audience for the expert debate, as well as being a part of it, so is subjected to profferings of apparently conclusive, final, once-and-for-all end-to-the-matter data from both sides. Such evidence is therefore directly contradicted, no matter how enthusiastically it is promoted.

The various media have a crucial role in communicating and hosting the debate, meaning that access to these arenas is a vital first step in securing effective participation in the debate. However, once this has been secured, other strategies are necessary to frame expert contributions and set one apart from another. The various strategic approaches in evidence in this debate are explored below.

Strategy 2: Independent Experts

The notion of the expert implies a specific area of interest and a commitment to the pure pursuit of knowledge about that subject. However, as has been shown, the demonopolisation of expertise has tainted this notion where it applies to the interested parties who make up specific risk debates. Instead of being engaged in this pure pursuit of knowledge, experts find that the concept of simple truth finding is insufficient to secure them definitive expert status. The commissioning of research *by* one or other side in the debate allows other debaters to raise questions about the research's quality and neutrality. As such, the objective, immutable nature of science is called into question at the same time as being held up as the ultimate in proof. Additional strategies are therefore used to promote and market this ostensibly objective product, because the experts themselves represent interests that taint their association with the science they promote. Despite this corruption of the idea of the expert-as-scientist, faith remains in the notion of the expert itself, and the appeal of finding and being able to claim the support of the mythical über-expert that is devoid of bias remains strong. If the audience is sceptical of the interpretations of the front-stage experts who are close to the reality they are trying to dictate, then perhaps they will believe some distant scientist who can be presented as being on a quest only to discover the truth. The vulnerability of statements to accusations of bias and partiality thus leads to an increased value being placed on the notion of independence. To use an independent source, therefore, is: '[t]o erase the emitter (the social subject engaged in the production of knowledge) … to naturalize the cognitive processes, to claim that knowledge somehow derives naturally from the essence of reality' (Leps 1992, 98).

The most sought-after supporting research is therefore that which was designed and carried out independently and subsequently found to support a particular viewpoint, or that which predated the emergence of the debate, as this can be presented as devoid of bias or self-interest. The attraction of using previous and/ or externally produced research is clear when the scepticism with which internally commissioned research is treated can be understood. Government-produced research was particularly vulnerable to accusations of bias, given that most of the research directly produced in order to contribute to the debate was made possible by government funds:

> I think there should be proper independent research. You know that report on speed cameras was published '*on behalf of the DfT*' and it was actually barking mad when you read through the detail of it. (Interview with Chief Constable 2)

> It's time they [the government] stopped commissioning reports designed to demonstrate that their flawed policies are working and started listening to their critics. (ABD Chairman, ABD 2004d)

> The ABD have called for a full independent audit of cameras as have the Conservative party. (ABD 2004a)

> There's clearly a pressing need for an independent audit of the Partnerships – not the internal reports we've seen to date. Asking the camera partnerships to audit themselves is quite bizarre. (ABD 2004e)

This criticism could, however, also be levelled at those opposed to speed camera use, who were described as producing '[s]elf-serving clap-trap' which promoted their own interests as drivers, not the 'reality' of the situation (Mary Williams, Chief Executive of Brake, BBC Breakfast News 2003).

One explanation for a contradictory finding is therefore that the seekers of truth were not in fact objective scientists but interested parties looking to create evidence that supported a predetermined position. The rational, objective promises of science are seen to have been corrupted by interest. Opponents who can be shown to be acting on self-interest and with ulterior motives can be shown to be acting in a less than rational (and hence scientifically flawed) manner. Instead, they are 'obsessed' with (Chief Constable 1; ABD 2004a) or 'addicted' to speed enforcement (ABD 2002; ABD 2009).

Research commissioned by a debating expert can therefore be dismissed the instant it is produced, with, in Beck's terms, '[t]he objections ... consumed *before* the results' (Beck 1992, 169). This pre-rejection of such internally produced science has led to an increased value being placed on the role of independent experts in producing an evidence base. Beyond this, the promotion of the involvement of experts has *itself* become a persuasive strategy:

> We deliberately used an academic institution to do it because we didn't want it to be seen as DfT's analysis. (Interview with consultant employed by DfT on first NSCP evaluation)

> The third-year report is done by someone from UCL, a Professor, so you can't argue. It has to be quite neutral, but people will still say, 'We don't believe the figures', but what else can you do? (Interview with senior Home Office official)

> The fact is the [attitude] survey results are compelling. They're done by an independent polling company, the people who do the elections. (Interview with Chief Constable 1)

It is also necessary that report authors themselves demonstrate their qualifications and their independence. As noted in the foreword to the RAC Foundation's 2010 review of evidence of speed camera effectiveness, the work is 'a thorough, independent, statistical evaluation' and its author 'has no axe to grind and no vested interest in the success of speed cameras. He is a respected academic with many years of analytical experience in this field; hence the reason the RAC Foundation approached him to undertake this study' (Glaister, in Allsop 2010, ii). In the quotation below, the idea of the 'compelling witness' further illustrates the ways in which the official agencies attempt to present their research and insulate their findings from accusation of bias or corruption:

We did get our ducks in a row. We commissioned the right research. We had very clever and convincing people to crunch the numbers, which is necessary because this is not just a tabloid debate. We've got the real movers and shakers, the thinkers, the intellectuals, the academics. And having Professor Heydecker's[1] name attached to the report is a million times more valuable than having mine on it, because although I know an awful lot about policing, I know next to nothing about statistics. Therefore I would not be a compelling witness. Professor Heydecker is. (Interview with Chief Constable 1)

Although authority is normally conveyed by the omission of a specific individual's name from such a report (Atkinson and Coffey 2002, 59), in this case it is necessary for the entire commissioning body to be deleted. Instead, the status claims (ibid., 61) are made through the use of titles which confer expertise and trustworthiness. In the case of the government's evaluations of the Safety Camera scheme, this involved the use of a 'Professor' and a 'Senior Consultant' and the near-complete elimination of any reference to the DfT or Home Office as involved parties.[2] Again, however, taking such a step back from the level of politics brings us no nearer to accessing this über-expert: '[t]his situation is rich with strategic opportunities. When political strategists still find it beneficial to legitimate social policy with reference to academic research, an unparalleled opportunity exists to select from the findings of different studies to justify polices' (Haggerty 2004b, 220).

Independent experts can, in this context of demonopolised expertise, be found to support opposing arguments. Although undoubtedly an 'unparalleled opportunity' in some senses, this situation also further undermines the notion of a single scientific truth, drawing attention to the fact that the same scientific criteria can apparently reach contradictory conclusions and do not, after all, provide the key to unshakeable truths about how the world simply *is*. The 'emitter' (Leps 1992, 98), initially erased by the notion of independence, is thus then reintroduced through the process of selecting the desired independent expert and presenting their findings in a convincing manner. The following example demonstrates with some irony what Haggerty (2004b, 220) has referred to as the unparalleled opportunity presented by the existence of multiple, conflicting expert views:

What we are going to do is to look at, perhaps do some research, an internal workshop. We've invited a few key academics, invitation only, doors already

---

1   Professor of Transport Studies at UCL and co-author of the two- and three-year evaluations of the NSCP.

2   While the core authorship of the 2003 and 2004 evaluation reports (Gains et al. 2003, 2004), commissioned by the government, is the same, the first is headed by the DfT logo. The 2004 report, however, features the UCL and PA Consulting logos and makes no reference to the DfT anywhere on the title page. The DfT is the main contact address for the 2003 report, while only the UCL and PA authors' contact details are given in the 2004 report.

closed, but it'll be people you've heard of, people like Steve Stradling, Claire Corbett and a couple of others. What we are trying to get from these academics is for them to have a slightly more learned discussion about what their understanding is of not just cameras but the whole speed management problem, what you can and can't do, what works and what doesn't, but on the assumption that we can't put cameras everywhere, there aren't enough cameras, what else can we do? (Interview with senior DfT official)

Academic reinforcement for the safety camera debate is, seemingly, now a matter of selecting the appropriate independent expert, inviting them in and then publicising the fact that you have them behind you, supporting what you say.

Strategy 3: Discrediting Opponents

If science is presumed to be able to offer conclusive proof, the existence of competing scientific findings needs to be explained by experts engaged in risk debates. This can be achieved, as has been shown, by suggesting that the author is biased and therefore looking for only one result to emerge from their research. However, in addition to this questioning of an opponent's objectivity, the strategy of questioning the *skills* and the *scientific ability* of the opponent's researchers also helps to explain away the fact that they, too, are claiming to have conclusive proof. If their abilities as a scientific expert can be questioned, then their right to claim that status can be thrown into doubt. The audience can then be encouraged to dismiss whatever they have to say as falling below the required standards of proof that the experts themselves have established.

Expertise is therefore not only created and established within the debate but attacked and devalued in equal measure. In a debate based around the truths obtainable via the use of scientific principles, an attack on the credibility of an opponent's scientific abilities, and thus evidence base, is a powerful method through which their voice as an effective expert can be undermined. The principle of science remains valid, as Beck assures us, so perhaps the only way of discrediting an apparently scientific finding is to question the skill with which those principles were applied. As Haggerty notes, however, 'we are expected to presume that the empirical work of the statistical debunkers is methodologically sound', and must trust it to be so in order for this tactic to be successful (2002, 97).

In the speed limit enforcement debate there is ample opportunity for allegations of the inappropriate use and generation of data. Experts in this debate explain away the apparently contradictory findings that the same criteria and methods of 'science' have apparently produced: 'They', too, have used science, but 'they' are interpreting and presenting it wrongly. In this case, this is presented as being through lack of skill, rather than through the deliberate contaminating bias and self-interest with which non-independent expertise is associated. In the following examples, opponents' skills (in these cases, those of the status experts and the scientific experts they have employed) are compared unfavourably to those possessed by school pupils:

> [S]ome parts [of the second NSCP evaluation] … if you'd been a sixth former submitting a piece of work like that you'd have been severely smacked! (Interview with Chief Constable 2[3])

> This argument is so full of holes it's hard to know where to begin attacking it … They begin by making four fundamental statistical errors which any A-level student should be familiar with. (ABD 1999a)

The comparison with school-level statistical mistakes not only makes the government (in this case) look naive and impossible to trust, but reinforces the idea that members of the audience need 'real' experts to make these observations for them. In the following examples, drawn from opposing sides of the debate, the scientific flaws in the opponent's work are made explicit to add to the persuasive power of the exercise:

> In the report it actually says that it's not possible to make 'before' and 'after' comparisons, but then they go on and construct some statistical model to try and do just that! It's all estimates, and they make the cardinal sin which I was told never to do, which is that you shouldn't use percentages of percentages. And yet they do! (Interview with Chief Constable 2)

> I've crossed swords with the ABD on some of their Mickey Mouse statistics and some of them are very easy to defeat, and most of the people spouting them don't understand them. Paul Smith [SafeSpeed] thinks he's a statistician. He's the guy who's got obsessed with regression to the mean, but he's using it to make a point that cannot validly be made from the statistics we've got. (Interview with Chief Constable 1)

It is not necessary for the audience for such statements to understand the phrase 'regression to the mean' or understand the reasons why percentages of percentages are not to be used. Rather, they are deployed as impressive-sounding and suitably 'expert'-sounding phrases designed to portray the issue as complex and scientific. Simply by offering convincing-sounding deconstructions of opponents' arguments, these experts are able to emphasise not only the complexity of the issue but the need for experts such as themselves to interpret it. Reliance upon one expert is thus increased via their offer to interpret and demystify the flawed expertise of another expert.

---

3   Chief Constable of an area that had chosen not to install speed cameras.

Strategy 4: Marginalising Opponents

To imply that a viewpoint, a position or a belief is marginalised is to imply that it contradicts what the majority of 'normal' people believe. To suggest that one view is held by right-thinking, logical people and another by quite the opposite is a potentially appealing strategy for any expert wishing to discredit and lessen public support for their opponents. To portray one's opponents as being supported only by irrational extremists is therefore to threaten anyone tempted to oppose your view with the same label. Rather than attack the science that opponents use, the idea is put forward, in this strategy, that their interpretations are in denial of the facts and as such held by only a few ill-informed, or even actually insane, people. Rather than just blaming corrupted self-interest, or the misguided application of scientific principles for the existence of contradictory research findings, people with opposing views are shown to be disinterested in the proper criteria of science, and as such can never be convinced of the error of their ways. In this way, the intransigence of the debate can further be explained by one's opponents' inability to even argue on the same terms. If one's opponents are irrational, then this explains their failure to be convinced up to now, even by the best possible evidence.

While appearing to be rather simple, unsophisticated insults, the following examples can also be understood as promoting the opposition view as, at best, ill-informed and, at worst, insane. Aligning oneself with people described as 'silly', 'bonkers', 'living at the edge of rational thought', 'a bit odd', 'the lunatic fringe' (Chief Constable 1) or, conversely, 'rabid', 'apoplectic', 'fanatics' (Mutch 2002[4]) is to align oneself with people who effectively exclude themselves from rational debate. Similarly, the description of the Association of British Drivers' evidence base as dependent on 'lonely factoids' (Slower Speeds Initiative 2001) suggests that the evidence itself is marginal and isolated. The author and *Guardian* columnist George Monbiot's description of the case against cameras as being promoted by 'cranks and quacks of all descriptions' (Monbiot 2005) achieves similar ends.

In the following examples, the peculiarity of the opposition is, it is implied, directly associated with their stance on the issue of speed cameras. Agreeing with them *is* considered to be an option, but one that is taken at the expense of being taken seriously in the debate:

> It's actually difficult now to be against cameras, unless you're a nutter, but there are a lot of nutters out there, of course … I think they've shot their bolt. I think they've revealed themselves for what they are. Some of these people do sound odd on the radio, they look like wide-eyed eccentrics. People think, 'Well, they're a bit odd this bunch'. (Interview with Chief Constable 1)

---

4   Ian Mutch writes for *StreetBiker*, a magazine for motorcyclists, and authored an article entitled 'The great speed debate: The nation awakens' (Mutch 2002).

> [The ABD's] statistics are simple self-serving claptrap that's been propagated
> by fringe groups who really aren't the road safety experts that know their stuff.
> (Mary Williams, Chief Executive of Brake, *BBC Breakfast News* 2003)

The implication that an opponent holds what is not only an irrational but a *marginal* viewpoint is therefore promoted. In a bid to marginalise the view of the opposition, the ABD has, for example, been challenged as to its representativeness through accusations that it lies about its membership numbers:

> It claims to the 'voice of the driver' which represents a groundswell of opinion
> among Britain's motorists. But the leading organization campaigning against
> speed cameras is dominated by a small core of libertarians and has routinely
> allowed its membership numbers to be exaggerated ... ABD's chairman, Brian
> Gregory, initially told *The Guardian* that the association had 9,000 members.
> But when challenged, the ABD lowered its claim to 2,256 paying subscribers
> and 3,775 'affiliate' members. (*The Guardian* 2004a)

In this example, an opponent of the ABD's position on speed cameras attempts to discredit the organisation by portraying it as untruthful. The key issue over which it is challenged is its membership, which can then be used as evidence of both its dishonesty *and* marginalisation. In line with the overarching scientific principles of the debate, attempts to marginalise the opposition are also supported with research evidence as to the *extent of* this marginalisation: 'We still have this rather strange minority of oddballs, most of them, ne'er-do-wells, and misfits and malcontents ... but we can show from Professor Stradling's work that they are extraordinarily small in numbers' (interview with Chief Constable 1).

Of course, in this debate the 'true' measure of an assertion of the minority status of an opponent is that someone (independent) has gone out and actually *counted* them. As such, even the apparently unscientific trading of derogatory statements has its own expertise underpinning it. Not only are alternative viewpoints presented as being marginalised, but they are *increasingly* so:

> We still have people attacking cameras, Motorists Against Detection and the
> Association of British Drivers, but they seem to be comprehensively losing
> the debate ... these people, I think, have expelled themselves from the debate,
> because they have been shown to be silly, and they do not command public
> support. (Interview with Chief Constable 1)

> I think the turning point for me came last September when Linda Lee Potter
> [columnist] in the *Mail*, the bastion of all bastions, came out and said cameras
> were a bad thing. And that to me is a real, it's a significant turning point. People
> are pushing away from them. (Interview with ABD spokesperson)

Things are changing. Some found it difficult but I think when you explain what we're trying to do and they see the success that we're having at reducing casualties where other places haven't, they start to come round to our way of thinking. (Interview with Chief Constable 2)

I think I'm beginning to take quite a bit of heart now that nobody's had a serious go at me for several months … Captain Gatso [Motorists Against Detection] scarcely raised his head in the last six months. Paul Smith [SafeSpeed] hasn't got any airtime anywhere recently, and the more odd letters they write in on 16 sheets of foolscap, the more they tend to disable themselves as an effective lobby group. So I think overall – at the risk of being smug or complacent and self-satisfied – I think it's working. I think we're winning … Anecdotally, people are on side. They are not rising in rebellion. People have stopped writing to the newspapers around here … I really do think the culture is shifting. But then there's always been a majority who are *for* this. (Interview with Chief Constable 1)

These examples, drawn from opposing sides of the debate, demonstrate that both are keen to promote the idea that they are winning. While unable to deny that the opposition exists, it can be presented as becoming increasingly isolated and abandoned by the public. The opposition's efforts to say exactly the same thing can even be presented as evidence that the other side is anxious and knows it is losing the battle for public opinion. As such, the idea that common sense has finally won out is reinforced. The public is invited to join with the majority on this issue. Its past misunderstandings are forgiven and it is welcomed on to the winning team – whichever that may be.

That this strategy is in use by both sides in this particular debate helps to explain why the debate itself has shown no signs of being resolved. As a strategy, it makes more sense from the point of view of those experts trying to employ it than it does from the point of view of the public on the receiving end of it. For the audience, subject to the same strategy being deployed from polar opposite stances about the same issue, there is no simple choice to be made. How is an individual to decide when faced with the same strategies being used to promote opposing viewpoints?

Strategy 5: Homogenising Potential Supporters

Predictably, the expert debaters also have a suggestion for how this decision can be made. The three strategies considered here under the heading of homogenising strategies rely on the audience's presumed desire to be part of a majority, and in that sense also draw on the fears of exclusion identified as underpinning the strategy of the marginalising discourses in evidence in this debate. These approaches also rely on a belief that this majority shares certain viewpoints about issues that can be exploited and used to others' advantage in the furtherance of their campaigns.

The three strategies considered here relate to the evocation of the idea of *a* commonsense viewpoint on speed enforcement, the use of a particularly emotive construct to exploit a lowest common denominator and, lastly, the utilisation of the very *idea* of 'science' as a homogenising strategy itself used in support of the science underpinning a particular argument.

These particular supplementary, supporting strategies rely on 'inductions' (Leps 1992, 45). The mobilisation of such terms in support of a particular interpretation relies on the presumed existence of shared norms and understandings which give certain concepts or statements a particularly strong referential meaning. Inductions do not require explanation and actually draw their power and meaning from this fact (ibid.). Of course, being the debate that it is, such unspoken logics are used in support of completely contradictory explanations and statistics.

The Commonsense Consensus

The exploitation of the notion of common sense is one way in which two similar appeals from different sides can hope to distinguish themselves. Rather than rely on the promises of the indisputable nature of science (promises that have failed to live up to their claims), experts can be seen to appeal to a much more intuitive understanding of the way the world works. In this way, experts can hope to recruit, persuade and cajole their audience into accepting their view of the world by virtue of claiming a sense of fit with what that audience intuitively knows to be true. A virtual risk or one perceivable only through science can be therefore be promoted as though it were a 'directly perceivable' risk issue and 'managed instinctively and intuitively' (Adams 2003, 87).

Strategies for both creating and exploiting homogeneity based on shared notions of common sense give the impression that the majority are on a particular side and thus exploit the audience member's presumed wish to be part of this, rather than one of the marginalised 'oddballs' (interview with Chief Constable 1) who disagree with them. In this sense, the exploitation of notions of homogeneity is closely paralleled with the opposite threat of marginalisation. No one, it is assumed, wants to be marginalised (uncommon) and non-sensical. If individual audience members find themselves disagreeing with a particular viewpoint promoted in this way, they will also find an attached implication that they disagree with the rest of rational society and, given that this is the case, have effectively excluded themselves from rational debate on the issue.

As with every other aspect of the debate, the public is subjected to similar appeals from both sides. Consequently, the congruence of either with what the individual intuitively feels about the issue can be seen as likely to promote that interpretation over one that seems to contradict what is 'common sense' to them. The task of the expert, therefore, is to promote their own interpretation in a way that stresses its own fit with common sense, and the way in which other interpretations do not. Thus, a sense of homogeneity and of shared values is as much *constructed* as it is pre-existing and exploited. In the following examples, the idea of a commonsense public opinion consensus is first created and then invoked in support of the particular statement:

> The motoring public realise that the argument around enforcement and fixed cameras isn't sound anyway, so the organisation [DfT] loses its integrity and credibility. (Interview with Chief Constable 2)

> Because the official policy on speed is risible to anyone who has given the subject any thought, pub talk across the land is hostile to all road safety policy. (ABD 2004d)

> It is now widely accepted that the switch from traditional traffic patrols to policing by camera has caused problems on the roads. (ABD 2003a)

> Clearly the public are rightly adamant that cameras not positioned at accident blackspots must go. (ABD 2003b)

> 'A 1 mph change in average speed causes a 5 per cent change in accidents': It is quoted *ad nauseam* in most Government anti speed publications. Now to any reasonable human being who knows anything about driving, this statement is clearly absurd. (ABD 1999a)

To disagree with the sentiments expressed in the last example is therefore to concede either to not being a rational human being or to knowing nothing about driving. Either admission effectively excludes you from being taken seriously within the debate, and the appeal of adopting the view suggested is therefore plain.

In the following examples, an intuitive 'warm' understanding of the world is used to reinforce the 'cold' facts of science, resulting in findings that are both logical *and* proven.

> We can demonstrate quantifiable and significant reductions in road speed, and anecdotally, which is probably at least as important because this is a dinner-circuit topic … it's a pub conversation topic. I now meet lots of people who say, 'It feels better on the roads, we can actually feel the difference. We are more relaxed.' And our valid statistical research does show the same thing, that offending has reduced, which is what we set out to achieve. The mood music is good … and mood music is at least as important as scientific evidence. (Interview with Chief Constable 1)

> The conclusion of the TRL report is therefore as statistically invalid as it is rationally absurd. (ABD 1999a)

> To suggest that reducing speed plays no part [in reducing road death] defies
> research and reality. (Gifford[5] 2004)

The perfect evidence base is clearly one that *makes sense* on both levels; the worst
therefore logically fails on both counts. Science, if it is to be most effective, needs
to be marketed as backing up what we already 'know' to be true, with this sense of
feeling right or wrong perhaps the only thing distinguishing between alternative
interpretations, given that all science is promoted with equal conviction. The
alternative, however, to this sense of intuitive 'rightness' suggests that the most
rigorously supported scientific finding will fail to convince if it goes against what
we feel we know to be true about ourselves and the world we live in. As has been
shown, the public is subjected to similar commonsense appeals from both sides of
the debate, and the congruence of either with what the individual intuitively feels
about the issue can be seen as likely to promote that interpretation over one that
seems to contradict what is common sense to them.

The 'goes without saying' logic of the use of common sense as a referent can
therefore be seen as erasing the idea of the emitter in a similar way to the use of the
independent scientist explored earlier. By appealing to notions that are presumed
to exist naturally among the audience, the sense that either side is acting out of
self-interest or bias can be bypassed in that it seems to be supported by self-evident
facts about the nature of reality. If the vagaries and contradictions of the debate
were removed and we thought independently about the issue, the experts would
like us to think that it is *their* interpretation that would seem to appeal intuitively
to us.

'Playing the Dead Child Card'

In a staged television debate, a Chief Constable who had chosen not to use fixed
speed cameras in his constabulary area was pitted against a representative of the
road safety charity Brake. The Chief Constable blamed 'misleading statistics' for
claims that his area had suffered from rising road casualty figures. He noted that,
in fact, government figures confirmed that his 'annual totals were still far below
that which we are required to achieve' (*BBC Breakfast News* 2003). The response
of the Brake spokesperson was as follows:

> [Cameras] have made an enormous difference on this road. This road has
> claimed the life of a two-year-old girl … If I was a bereaved mother in [County]
> and had lost my child to speeding, I would call up the Chief Constable and say
> 'You are in breach of your duty of care to your community'. (Ibid.)

---

5   Rob Gifford is Executive Director of the Parliamentary Advisory Council for
Transport Safety.

This debate was then described to me in an interview with a third party who sympathised with the Chief Constable:

> Unfortunately he brought out cold hard statistics, which are OK to a point, but she had a story about a killed child which trumps anything. I've heard it referred to by people outside the ABD as 'playing the dead child card' which I think is a bit cynical, but it's very effective. (Interview with ABD spokesperson)

The death of a child is a powerful image that can be used to both reinforce and undermine more quantitative input into the debate. It is in stark contrast to the 'more abstract, formal, methodical and inaccessible knowledge' that constitutes expert discourse (Ericson 1994, 156). If nothing else is certain in this debate, experts can presumably rely on the fact that most people will react to the death of a child. This approach to the debate has been described, with some cynicism, by one author as 'the demonization of speed per se proceeding with the assistance of gruesome images and traumatic tales' (Mutch 2002).

The deliberate and strategic use of 'the dead child card' is illustrated powerfully by an example from the arena of road safety education. The DfT's 'It's 30 for a reason' education and publicity campaign was explicitly underpinned by this logic, as the press release about its launch testifies:

> Two key strands were identified during creative development research carried out to determine the best ways to gain the attention and support of the general public. One, the emotional effect of hitting a child whilst speeding, and secondly, the use of statistics and science to support the scenario presented ... The responses from research identified the need to combine the emotional with the rational in order to avoid only a short-term 'tear jerk' effect. The commercial has been designed specifically to marry up the emotional versus the rational aspects to target the 70 per cent of the population that speed. (DfT 2005d)

This example also demonstrates the extent to which the debate among expert interpretations necessitates secondary research into the best ways of communicating the primary scientific research in order to obtain maximum impact. 'Research' has shown that the combination of data and drama has the greatest persuasive effect and consequently that is the approach adopted. The fact that the government has researched this phenomenon is then also considered to be worth revealing, with the result that even *this* strategy has its own evidence base with which to impress the audience: Research reveals one course of action to be the most viable, research recommends the best way of achieving this, and research suggests that the existence of research into the research has persuasive potential in itself.

The 'card' does not stand alone, but is always employed in the furtherance of a particular interpretation of the causal chains at work in relation to road safety, to support a specific stance as the only way of preventing the agreed-upon undesirable and tragic outcome. The death of a child is seemingly, to revert back to the core issues at the heart of this debate, a vivid and tangible illustration of the failure of a particular interpretation of the causal processes at work. It serves as a painful and emotive reminder of the consequences of getting these interpretations wrong. As a discrediting tactic, it is extremely effective and illustrates dramatically that the context is not solely dominated by scientific discourses. When faced with this kind of emotive and populist challenge, science is left wanting.

Although often centring on the death of a child, this strategy can, in fact, deploy any death to persuasive effect. Death is the visible manifestation of the harm that underpins the debate and which both sides claim to be in the business of preventing. As a tactic, the notion of *caused* rather than *prevented* deaths is therefore used by both sides of the debate to 'trump' the argument of the other, as in the following example from a debate hosted by the BBC:

> *ABD spokesman*: If you look at the statistics, the trend up until 1993 was a 7 per cent fall in fatalities. Since 1993, when the speed cameras and specifically the Speed Kills policy came in, it is down to 2.9 per cent. That's an incredible increase [in] the number of fatalities. If that trend (pre-1993) had continued from 1993 to date there would be 5,000 fewer deaths. The Speed Kills policy has made road safety worse.

> *Brake spokeswoman*: If you drive at 30 mph and you hit a child – and children will run out into the road chasing a ball – there is a 50 per cent chance that that child will die … [ABD spokesman] is talking absolute nonsense and I would challenge him to say these words to Sarah Turner whose 12-year-old son was killed crossing the road outside his school. (BBC News Online 2003d)

Via this approach, an opponent can thus be accused of failing with regard to the one legitimate and permissible preventative aim acknowledged in the debate, with this failure brought into play for both sides as a tangible and emotive example of their failure to protect innocent victims.

As a debating strategy, the notion of a caused death is at once powerful, emotive and extremely difficult to argue against. Such a strategy forces an opponent to defend him or herself and to attempt to explain how the holding of a particular view does not inevitably make them in favour of killing people. The individual so accused is thus cast in the role of violent offender, a position they then have to argue their way out of in order to resume the role of credible expert.

The Idea of 'Science'

The final version of the homogenising discourse considered here is the use of preconstructs from the natural sciences to attempt to unite the audience behind some of the most basic and fundamental scientific principles on which the world operates. Through the use of such preconstructs, 'the aim is less to prove the truth of the argument than to stage it: to represent its validity by an illuminating example which will strike the receiver as so obvious that it will provoke an immediate recognition rather than a reasoned acceptance' (Leps 1992, 51). This equation with simplicity is belied by the fact that both sides are able to use the strategy from opposing positions, with the existence of the debate itself proving that matters are not as simple as each side attempting to suggest. The examples drawn on for this purpose must be naturalised, commonsense understandings of science in order to be effective, and must invoke in their audience a faith in science and an intuitive belief in its ability to explain how the world works. The strategy therefore relies on making a connection between one of these fundamental principles and the particular causal interpretation being promoted. Three examples of this strategy are considered and dissected below.

In the following example, the use of the concept of 'simple Newtonian physics' in support of the Chief Constable's favoured causal interpretations is intended to imply that, stripped of all the personalities and politics of the debate, we are left with the simple incontrovertible fact that cars crash because they are moving. He noted that '[i]t's simple Newtonian physics! If your car is not moving, you cannot crash into somebody and kill them' (interview with Chief Constable 1).

This is a persuasive strategy rather than a simple scientific observation, however, because it is used to imply that the official reaction in the form of speed cameras is similarly logical and appropriate. Although it is indeed undeniable that, without movement, cars will not crash into each other, the use of speed cameras to intervene to prevent this happening cannot be justified using the same logic. Cameras are one of many potential interventions that reduce the speeds at which vehicles are travelling and in doing so reduce the speeds at which they would crash if on the same path. They are, however, only one option for achieving this result and they do not intervene in a manner that rules out the potential for crashes to occur. Similarly, although speed is not the only cause of a crash, the implication of the Newtonian reference is that cameras will prevent all crashes by preventing 'speed'. This approach presents the use of speed cameras as a similarly logical response and allows it to benefit from the indisputable scientific logic of 'simple Newtonian physics'.

In a second example, 'evolution' is the scientific justification appropriated in support of the NSCP: 'You can be religious and cope with evolution, but we have people who try to reject the concept, and it just makes you bonkers because evolution is a fact, and it is a fact that safety cameras do what they say' (interview with Chief Constable 1). This author implies that the same amount and quality of evidence exists in support of speed camera effectiveness as it does in support of the theory of evolution. Such is the way it is presented that this expert almost appears to be claiming that evidence of one is also evidence of the other. The author notes that if 'even' religious people can accept the scientific principles of evolution, no one should be capable of rejecting the scientific premises on which speed cameras are justified. Those who reject the logic of the speed camera as an intervention are therefore compared to individuals who deny evolution and who are also, according to this Chief Constable, 'bonkers'. To deny the 'fact' that cameras 'do what they say' is therefore to exclude oneself from rational debate, to marginalise oneself and to declare oneself unconvinced (and, furthermore, unconvinc*able*) by scientific 'fact'.

The final quotation in which this strategy is used relates to one Chief Constable's admission, reported by the Association of British Drivers, 'that some speed limits were "barmy"' (ABD 2005c). They claimed: 'You don't have to be Einstein to work out that his position on cameras became untenable as soon as he uttered those words' (ibid.). The preconstruct of Einstein's intelligence is used to imply that the whole policy of the NSCP and speed camera use more generally is undermined by this admission in relation to *some* limits. The fact that none of these 'barmy' limits may be actively enforced by speed cameras and as such are unconnected to the validity of the policy is circumvented by the implication that the connection is obvious and does not require a genius to work it out. Given that you do not, according to this example, have to be Einstein to work out what this means for the policy, we are encouraged to believe that even unintelligent people can reach the same conclusion as this author. Those who do not agree are, therefore, presumably even worse than 'unintelligent'. To avoid such an implication, therefore, the reader is encouraged to agree with the author and, by extension, their interpretation of the causal interpretations at work in relation to the issue of speed.

Each of the examples considered above invites the audience to make a choice. This choice is between aligning themselves with some of the greatest scientific thinkers in history (and/or their scientific discoveries) and professing to know better than them. Effectively, a situation is created whereby to disagree with the statement being put forward is to question gravity or the theory of relativity. As such, this strategy is linked to a wider use of the notion of common sense to imply or attempt to construct a basis in the understood reality of the way the world works.

These tactics demonstrate how the debate has gone full circle, with commonsense taken-for-granted scientific principles being used to reinforce marketing strategies that are in *themselves* designed to promote science. The appeal of preconstructs such as the intelligence of Einstein is, however, a strategy that actually bears no relation to the science it is used to promote. In each case it actually conceals a scientific error, a causal leap that cannot be justified. Ultimately, our faith in science is exploited as a strategy for the promotion of science. This is achieved via the idea of common sense, which attempts to imbue each expert's science with the same intuitive qualities possessed by invoking the names of Newton and Einstein – familiar to all the audience as 'real' and undisputed scientific experts.

The Expert Marketplace

This chapter also considers a second consequence of the demonopolisation of expertise, this time in relation to the impact of the debate upon its audience. It considers that the experts' apparent inability to agree may be partly responsible for an apparent independence of the public from such expert sources. As such, the possession of expertise can be said to have been *democratised*. Rather than demonopolisation resulting in an expertise *vacuum*, a situation is produced in which everyone can, and does, claim a degree of expert status in relation to speed and speed enforcement. That the traditional expert voices considered previously have failed to recognise this fact is, it will be suggested, a further explanation for the debate around speed limit enforcement.

The Democratisation of Expertise

Beck suggests that the demonopolisation of expertise has, in addition to necessitating the presentation and promotion of science, resulted in the 'equalization' of experts and non-experts. However, he generally confines his evidence of this process to the growth of pressure groups of injured parties, using the notion of 'democratization' only in relation to the spread of collective action outside of traditional democratic systems of representation (1992, 191). The democratisation process that will be discussed here, however, extends this to include both the challenging of expert interpretations by ordinary, lay members of the public, and the partial emancipation of the public from such sources of expertise altogether. In both senses, therefore, the distinction between the expert and the non-expert is challenged. As such, the democratisation process with which this work is concerned is more akin to that described as 'empowerment' of individuals by Giddens (1991, 142), or characterised by Young as an 'increasing ... level of scrutiny and demand' from the public about the activities of those who seek to control, direct or regulate their behaviour (1999, 78).

In Beck's sense, however, the democratised general public has been presented as challenging expertise in only a limited sense, concerned with holding experts accountable when they fail, not in any sense challenging them as alternative sources of knowledge about risks. This limited sense of equalisation is explained by virtue of the fact that we cannot 'know' the risks to which we are exposed 'so long as to know means to have consciously experienced' (ibid., 72). However, there are some risk debates that are more accessible to the public than others, as Giddens notes:

> The sceptical, mutable nature of science was for a long time insulated from the wider public domain – an insulation which persisted as long as science and technology were relatively restricted in their effects on everyday life. Today, we are all in regular and routine contact with these traits of scientific innovation. (1999, 1)

In cases such as the simmering virtual risk issues of global warming, carcinogenic habits or even the dangers of eating British beef, the reality of the dangers is to some degree still inaccessible to the general public. Their consequences and the processes via which they occur are not overt or easily grasped by the lay public, who are thus still reliant to a greater degree on the explanations of scientific experts. In this sense, some risk debates may be more susceptible than others to the image of the public as empty vessels envisaged by the experts, perhaps accurately described by O'Malley in terms of '[l]ay people offering their "uncertain" estimations ... readily subjected to the more systematic and "objective" risk-based calculations of expertise' (2008, 464). However, risks that relate to technologies which *have* crossed over into the realms of 'regular and routine contact' may appear to require less expert interpretation (Giddens 1999, 1). As Roberts notes, in relation to the use of experts in court, 'expert evidence is admissible whenever – but only in so far as – it will help the jury to determine a question which is beyond the knowledge and experience of ordinary people' (2002, 259). The following analysis will consider how, in the specific context of the debate around speed enforcement, the general public, rather than being passive recipients of the expert debate, are showing signs of expertise themselves. In this way, this particular issue is considered to be less suited to expert advice given that it relates to an issue that 'ordinary people' do not necessarily consider *is* 'beyond their knowledge and experience'. As such, it is proposed, the notion of expertise has itself become democratised, and has become so partly as a response to the absence of definitive or conclusive expert advice:

> Individuals are thus expected to engage in the risk management of everyday life in the context of escalating competition between experts and quasi-experts. Into this destabilisation of trust in expert knowledge individuals, to some extent, but only in a limited number of fields, seek to acquire a degree of expertise for themselves (Hunt 2003, 171).

Speed limit enforcement and its surrounding issues of regulation and restriction have, this section will propose, become one of those fields in which formerly lay members of the public *have* gained expertise that enables them to manage risk in the absence of consistent expert advice. Such knowledge seeking is perhaps to be expected given that this is one everyday risk issue with a direct impact on members of the public and for which they are also held responsible. Drivers, it is suggested, have a vested interest in participating in, and contributing to, this debate given that they are directly and differentially affected by the different outcomes that are proposed by different experts. Members of the public are faced with the same strategies of persuasion, deployed with equal conviction by both sides, in the promotion of a science that it is assumed they are unable to judge, yet are compelled to make a choice and engage with the debate in some way: 'We don't know and can't know – yet all of us, as consumers, have to respond in some way or another to this unstable and complex framework of scientific claims and counterclaims' (Giddens 1998, 24). To this degree, the competition between experts can be seen to have contributed to the undermining of the role of the expert in the formation of public opinions about issues, given that it does not offer the capacity to provide the answers that the public seeks: 'We come to recognize through experience that we can no longer rely on experts to guide us in the choices that we make and are forced to make decisions in the light of conflicting information. Science is losing its traditional role as expert adviser' (Franklin 1998, 5).

The following analysis considers the way in which public talk displays a degree of expertise about, and critical engagement with, the issues of speed and speed enforcement. It also considers how such lay experts, much like their traditional expert counterparts, are also concerned with establishing their expert credentials when they take part in the debate with other lay experts.

The Active Sourcing of Expertise

Rather than passive recipients of the debate, members of the general public demonstrate a willingness and an ability to actively obtain expertise from a variety of sources, including, but not limited to, traditional 'official' experts. They then show evidence of using that expertise in the formation of their own opinions about the issue, using it to reinforce and support the opinions they then choose to share. The examples used in this section feature ordinary motorists, not members of pressure or interest groups, who cite a wide range of sources to which they have turned to form or support an opinion on speed enforcement.

In the face of contradictory expert advice, the act of opening a newspaper, turning on the television or surfing the internet has become a personal research project, with a multitude of experts competing to provide the researcher with evidence. The sources from which expertise is apparently drawn by the public therefore provide further evidence of the demonopolisation of expertise and its displacement from the traditional arenas for debate into new and varied locations:

> You know they can't do you for it if the camera's not been checked that day, don't you? I read that on the internet. They have to validate their cameras every day, and if they don't, then you can appeal. (Female, mid 30s, Speed Awareness Course)

> If you don't sign the form [Notice of Intended Prosecution], they can't use it as evidence against you. I saw this article on a website that said that you can send the form in to the court like they tell you to, but if you don't sign it, then they have to get back in touch, and if it takes too long, then you get away with it. (Male, mid 40s, convicted driver focus group)

> These cameras are sufficiently 'clever' for want of a better word that they can set the parameters for different types of vehicle, different speeds for different types, you see? So if I'm following you in your car, it might get you not me. But they were doing them at 42 mph when the construction and use requirements are that they give you 10 per cent plus two ... I read this in the Road Haulage Association newsletter and it was at the meeting four or five weeks ago, and I paid attention because it was pertinent to my present situation if you like! (Male, early 60s, professional driver focus group)

Steve: I know the law, I've read up on it, but it's a case of having to know it really, being self-employed and with everything being in my name. Yeah, I make it my business to know, because I don't want to be getting caught out by a lack of knowledge or information

HW: You say you've read up on it. Where do you get your information from?

Steve: From the internet, so from lawyers, other people like that who have looked into these things, and from magazines, TV programmes, friends who have been through it. (Male, 60s, professional driver focus group)

There's a camera by Morrisons and the white lines were covered up by new tarmac some six months ago. So that can't be a real one, I mean a working one, as you need the lines for it to be legal. It's like a second-level check to back up what the radar said. A copper friend told me that. So that one's safe. (Female, early 30s, Speed Awareness Course)

As some of these quotations demonstrate, the internet is one valuable source of information for members of the public with questions and in need of answers, but active sourcing from a variety of locations is clear. The experts profiled previously are right to assume that there is a receptive audience for the expertise that they produce, yet their use of promotional tactics to communicate this data suggests that they make incorrect assumptions about the uncritical nature of their audience. When drivers have said that they have used the internet as a source of information, it is clear that they do not simply visit the websites of the DfT or local Safety Camera Partnership, but are open to a variety of sources from a variety of authors, the extent of which can be seen simply by Googling 'speed cameras' or a similar phrase. A review of internet-based discussion forums also reveals that the dissection of and debate about the official expertise supporting speed cameras is a popular topic, with some discussions involving fairly detailed and critical engagement with the material. Clearly, audience members are more active than might be assumed in terms of accessing expertise that they consider useful, and clearly 'traditional' experts are not the only source of information in this demonopolised expert context.

The Use of Personal Experience in the Formulation of an Expert Persona

In addition to the externally sourced research referenced by members of the audience for the speed camera debate, some individuals have placed reliance on more home-grown 'research' into the danger (or perceived lack of danger) associated with exceeding the speed limit. By constructing expertise based around personal experience, drivers are able to present first-hand, observed accounts of the causal processes relating to speed in which they have personally participated. Such explanations can therefore appeal on a very basic level, as apparently simple and therefore commonsense observations of what happens (or demonstrably fails to happen) when a speed limit is exceeded. Although those traditionally understood as experts have shown an awareness of the power of common sense as a tool to be exploited, they have done so as though it is something that can be created and shaped by the ways *they* talk about it. Experts using these strategies appear to think that it is persuasive simply to suggest that *their* approach tallies with common sense in a way that their opponent's does not. In this debate the audience *does* consult its common sense, but in way that *it* controls, not the experts. The subsequent analysis will demonstrate that, in the case of speeding, the assumed ignorance and exploitability of the audience is a very significant error on the part of the various experts discussed so far. Beck, too, assumes that in the face of conflicting information and the undermining of the notion of the expert, the rest of us 'are left panting in ignorance' as technology remains forever ahead of our understanding (Beck 1998, 13). However, where a risk activity is undertaken by the majority of the adult population, and as such is an activity about which many people have direct experience, it will obtain immediacy and a presence that is experienced on a more 'real' level than many other risks.

In this sense, driving-related risks differ from other risks, such as the BSE or passive smoking examples often cited in the risk literature. Individuals are able to make their own connections and apply their own interpretations of the effects of speed based on a stock of knowledge built up through lived experience. They are not issues about which individuals are solely reliant on 'the crumbs of information that fall from the planning tables of technological sub-politics' (Beck 1992, 223), but ones about which people can claim a degree of self-obtained knowledge.

The importance of this self-acquired expertise is reinforced by research that suggests that, where alternative sources of information are available, those interpretations offered by experts lose their strength (Druckman and Nelson 2003, 729). The less the public audience knows about the subject, the more they are free to accept what they are told. However, when they possess some degree of experience of the context in question, then it is logical to expect that expertise that fails to tally with their own experience will not be adopted uncritically. As such, the experts in the speed enforcement debate are preaching to an audience that comes to the debate armed with its own preconceptions about processes and outcomes, preconceptions based on the individual's very own direct experience of the causal chains operating in this risk context.

The successful completion (sometimes at speeds in excess of the posted speed limits) of thousands upon thousands of journeys by car serves to undermine the causal implication, contained within speed enforcement policy, that exceeding the speed limit is somehow inevitably and unavoidably linked to the chances of having a road crash. Lived experience therefore appears to directly challenge the causal interpretations favoured by the authorities, which appear to fail because they were only ever probability statements in the first place. Their inability to be deterministic therefore means that they can be rejected even as probability statements. As Garland observes, in an uncertain world, '[p]ast experience is our best predictor of future events' (2003, 53). These connections (or the lack of them) between behaviour and outcome can be experienced on a daily basis by millions of individuals in an immediate and real sense, and do not seem to involve the complex, delayed or indirect causal chains present in other risk debates. We may need to wait for 10 or 15 years to determine whether or not eating beef at a certain point in history was ultimately a risky thing to do (Giddens 1999, 2), but we do not need to wait at all to reach the conclusion that exceeding a speed limit on a specific occasion brought about no consequences whatsoever for our immediate health. Nor will it lie dormant to resurface at some undetermined later date. Alternatively, we are also able to appreciate at first hand the locations and situations in which we feel that we *are* posing a risk, or that someone or something is likely to experience us as a risk.

In the context of speeding, additionally, the impact of experience on behaviour has been noted. Corbett attributes the apparent failure of drivers to accept the 'real' level of risk caused by their speeding to the fact that harm is 'only very infrequently negatively reenforced' (2003, 105). The chances of any one individual actually suffering a speed-related crash are so small as to effectively neutralise the large-scale, generalised research findings about speed and danger. Effectively, this means that, by driving at excessive speeds and suffering no ill effects, the official messages about speed can be obliterated by the reality of experience. Although road death is a significant cause of death for some demographic groups in particular, it is notoriously underestimated by drivers. This is confirmed by a Brake survey that found that three in four drivers underestimated their chance of dying in a road crash: 'less than a quarter of motorists realise they stand a one in two hundred risk of dying at the wheel' (*Daily Express* 2001, 30; and see McKenna 1993).

In constructing their own expert identity, the following drivers combine their status as persistent speeders with observations about their lack of crash involvement. These are set within the context of significant experience to promote the credibility of their viewpoint:

> I've never had an accident in 34 years of driving and this is the first time I've ever been caught. (Male driver, early 60s, experienced driver focus group)

> I'm not going to change the habits of a lifetime when I've never been involved in an accident. I speed quite a lot, when I am in a hurry and – touch wood – that says to me that I'm quite good at it! I mean, I'm good at knowing when not to do it. (Male, late 30s, convicted driver focus group)

It is noteworthy that the quantification of driving experience was a particularly common way of creating an identity among members of the public engaged in the debate. In the absence of any other credentials, the claim to have 20, 30 or more years' driving experience or to 'driv[e] over 600 miles in any given week' (male professional driver, mid 40s, Speed Awareness Course) qualifies a contributor to speak with some degree of expertise. This was another strategy designed to improve the credibility of a statement on the issues under discussion. The following drivers, now in their late 50s, chose to introduce themselves in the focus group context with a description of their extensive expertise as drivers:

> *Dave*: We've got a lot of driving history between us. I passed my test three months after my twenty-first birthday and I've been doing it ever since. Rigids, artics, 32 tonne artics, the lot, and two years on 38 tonners. Then I've done all the training courses, the lot, and worked on that ever since. Every day of every year since 21.

> *Pete*: Same with me. Just after my twenty-first [birthday], so 30 years ago, and I've done all aspects. (Professional driver focus group)

Driving experience, in this sense used as a way of communicating a degree of expertise on the matter under discussion, forms a 'key category entitlement which participants claim to enable them to make authoritative statements' (Hine 2000, 129).

In the following examples, members of the driving public also demonstrate that they use their own familiarity with and experience of the policy itself in forming judgements about it. Their experience of interacting with the policy and their knowledge of the issue it relates to combine to form their expertise in relation to 'how things really are' out on the road. Specific instances of interaction with the technology are given as examples of situated, observed expertise about 'how it really works' which contribute to the formation of opinions in favour of or against the scheme. In the following example, such experience is used to construct expertise that promotes the idea of cameras as a risk-*producing* rather than risk-*reducing* intervention:

> We've got a camera right outside our depot – it's 40 mph outside where our depot is and 60 mph either side of it – and this bloke came through speeding, went past the camera, it was dark, the camera went off. It flashed straight through his rear windscreen, put him off and he left the road and hit a wall. The government tell drivers, 'Don't flash your lights in the back of someone else's car', and they put up cameras with flashes on them! And that's something I personally have seen so I know it does happen. (Male professional driver, late 50s)

> *Dave*: The only trouble with static cameras is everyone knows where they are and they'll make sure they're doing 40 mph for that 20 feet. But they don't know what the speed limits are. That's why people will drive to a camera and brake and then accelerate afterwards. That can create a very dangerous situation. Remember what happened to Ted?

Pete: Yeah! He very nearly ran into the back of a motorist doing exactly that. Both travelling, Ted reckons, about 55 mph, speed limit was 60. The guy comes up, sees the camera late and just [sound of hands clapping together] straight on the brakes. That's what cameras do. (Professional driver focus group)

Qualitative research carried out by Blincoe et al. (2006) confirms that such views about the accident-*causing* potential of speed cameras are quite widely held, particularly among those drivers termed 'manipulators', who themselves modified their speeds at camera locations before resuming their usual speeding behaviour.

The participants in this study, as well as those surveyed by Blincoe et al., have observed the reality of road use (as it applies to them, so far) for themselves and have used it to shape their views about the scheme. The drivers quoted above are able to view their observations as small-scale research projects, as a result of which they feel they are equipped to assess the issues of the debate for themselves. Such experience and application is, furthermore, considered to be necessary to equip the individual with the necessary information to function as drivers. In the following example, the formulation of expertise is described in some detail:

Ray: We used to watch people. I'd say to him [learner driver], 'See those cars ahead of you? What is that car that's overtaking you going to do when he comes in in front of you?' and he says, 'I dunno', and I says, 'That car is going to come in front of you and he's going to slap his brakes on', and I can guarantee you that every time I did it, every time, that happened. And they'd say, 'How'd you know?' and I'd say, 'Because I'm looking at him, and as he goes past he's going to spot that camera, and brake right in front of you'. They hammer it at maximum and brake for the cameras. But they don't just brake to that speed …

Dave: They don't know what it is for a start, it's just a reaction, they just brake.

Pete: … we did a bit of a 'survey' on it, if you like, and believe it or not, the worst ones are the women with children in the back. (Professional driver focus group)

In some cases, personal experience is used to supplement and support information purposively gleaned from other sources. The following quotation is cited at length to demonstrate the extent of democratised expertise shown by this particular individual, an individual who *combines* the two modes explored here, holding expert discourse up to the critique of personal experience to reach an ultimate view on speed enforcement:

I read *The Guardian* regularly and the others more infrequently. *The Telegraph* has quite a lot of motoring in it on a Saturday and I read that, oh and motorcycle magazines, and – joking apart – there is some very good information in that. And my stance, from what I've read and my opinion formed from those sources, is that speed cameras is an issue very largely to do with revenue and only partly to do with road safety. I understand that the majority of the most dangerous roads by casualties per mile have a lot less speed cameras on them than the ones that may or may not be revenue-generating. I'm extraordinarily sceptical, let's put it that way, because I travel the roads over the moors and they claim that between four and five bike deaths happen per year and the numbers of speed cameras on those roads altogether is three, yet on the A34 into Stafford there are 32 ... So even though I'm aware that there's a road safety element to some extent, I'd put that at 10 per cent with 90 per cent as revenue generation. So all these things give you a consciousness of what you think and what you are about and what your opinions are. It may be badly founded, or on the wrong documentary evidence, but the media obviously slants some of that as well. Plus chatting with people, who give you a different perspective on what they get up to. (Male, early 60s, experienced driver focus group)

As such, when experts appeal to 'common sense' to promote their viewpoint, they are not drawing on a previously unformed feeling. As noted above, a common strategy has been to present scientific findings as *doubly evidenced* both by scientific research and by their sense of fit with more intuitive notions of how the world works. Hands-on experience of the issue at the core of the debate means that members of the public are not empty vessels ripe for shaping by appeals from the experts or susceptible to being *told* that something is a commonsense interpretation of a sequence of events. In this sense, attempts to mould the common sense of the audience fail to take account of that audience's own experience and level of expertise in relation to the issue and, unsurprisingly, fail when that appeal fails to tally with what the audience feel it knows and understand of the issue. The counter-argument, of course, means that where an expert interpretation *does* fit with what the audience knows of the issue, it succeeds in being doubly evidenced and is more likely to be adopted as a convincing explanation of how the world, in relation to speed enforcement, 'really works'.

Conclusion

At the heart of this debate lies a partial misunderstanding of the relationship between expert and audience, a misunderstanding that helps to explain its intransigence. The contradictory science that is a feature of this demonopolised context has created a need for marketing techniques that attempt to promote a particular stance as the 'truth' with a view to achieving definitive expert status for each expert. Such strategies have attempted to exploit the assumed ignorance of the audience by promoting the independence of one interpretation, attacking the merits and scientific ability of another, threatening the audience via marginalisation strategies and cajoling them via homogenising strategies. The increased effort necessitated to promote and explain supposedly self-evident and self-explanatory research is, however, being exerted for the benefit of an audience that actually sees itself as increasingly independent of experts. This research has suggested that, rather than casting about in search of an opinion on this issue which they can simply adopt as their own, members of the public bring to the debate a certain level of knowledge produced as a result of their own experiments conducted in the course of their daily lives. When the driving public does engage with the traditional-level scientific expertise, it is therefore as a self-styled expert source in its own right, with its own preconceptions and beliefs about the 'correct' causal interpretations that apply in relation to the issue of speed. Scientific expertise, as such, *is* used where it reinforces existing beliefs, but will not be accepted if it fails to tally with the lived experiences of its audience.

That this 'expertisation' of lay groups has taken place is understandable, given that the debate is not only an abstract quarrel among self-appointed experts but a discussion that has real consequences in terms of its impact on people at an everyday level. It is these consequences and the way they are experienced and understood by the newly 'expertised' audience for the debate that form the subject of the next chapter.

Chapter 5
Respectability, Responsibility and Resistance[1]

This chapter considers both how speed limit enforcement is experienced by those who are subject to it and how that experience is shaped and influenced by its location within a society increasingly preoccupied with questions of risk. It considers the meaning and significance of criminalisation as a method of problematising behaviours that have been publicly linked, though not unanimously, to an increase in risk.

The preceding analysis of the debate that exists among experts provides an opportunity for those implicated as instigators of risk to reject the official interpretations of the processes whereby risk results on the roads and to select other alternative explanations. This chapter looks at why some of these implicated instigators might be tempted to take up that opportunity, again considered within the overarching risk context that has been identified.

Giddens notes that '[w]here consequential decisions are concerned, individuals are often stimulated to generate increased mastery of the circumstances they confront' (1991, 143). This chapter considers why speed limit enforcement has become just such a consequential issue for many people. It will be suggested that the regulation of speed, although justified on the grounds of risk, is experienced *as a form of risk in itself* in that it has negative consequences for the individuals implicated as instigators by the traditional defining experts of the previous chapter. This chapter suggests reasons why those on the receiving end of the policy might be tempted to adopt risk interpretations that allow them to excuse their behaviour and oppose the regulation of it, and explores the methods via which this is achieved.

Identity and Risk

For those who view society as not just increasingly concerned with risk but as an increasingly individualised and reflexive 'risk society', notions of identity based around the types of collective labels and identities that served well in the past are

---

1   Material from this chapter has also been published as Wells, H. (2007). Risk, respectability and responsibilisation: Unintended driver responses to speed limit enforcement. *Internet Journal of Criminology*, and as Wells, H. and Wills, D. (2009) Individualism and identity: Resistance to speed cameras in the UK. *Surveillance and Society* 6(3).

devalued and become less dependable and durable. The individual is freed from 'inherited identity' (Bauman 1997, 20) but instead has to construct and defend their own created version of the self:

> The general thrust of the theory of individualization is that 'given' forms of collective identity have been eroded and are being supplanted by more 'open' practices of personal choice and reflexivity. Under the risk regime, rather than life trajectories being governed by the ties of family, class, ethnicity and gender, 'do-it-yourself' biographies become the prevalent form of cultural determination. (Mythen 2005, 132)

Ericson and Haggerty suggest that many of the 'new' identities with which people increasingly identify are based around the categories to which they are assigned by agents such as the police: 'People identify their selves with their institutional categories of risk and the differentiated needs those categories foster. Individual identity is confirmed within the classification schemes and expert knowledges of institutions' (1997, 257). They note, however, that such categorisations 'create their own risks of deselection, marginalization, and exclusion' (ibid., 198). These are consequences, it might be suggested, that would actually discourage individuals from uncritically adopting the identities that 'risk' constructs for them. Instead, they may be attracted to alternative identity categories that allow such negative consequences to be resisted.

Risk-Based Speed Limit Enforcement and Threats to 'Respectability'

This section considers how some drivers have adopted specific identities in relation to speed limit enforcement. Rather than categories such as age, gender or class being significant, such identities relate instead to the individual's 'respectability' and centre around 'law-abiding' and 'moral' activities which situate the individual securely within 'the majority'. These identities are challenged by the way in which enforcement has been pursued under the NSCP and continue to be enforced after its cessation.

In order to explore the usefulness of the suggestions (above) in explaining the speed limit enforcement debate, this research has deliberately used methodologies that have allowed participants to construct their own do-it-yourself identities (Mythen 2005, 132) and classifications, rather than assigning them to more traditional categories based around class, gender or race. In doing so it has allowed identities to emerge which appear instead to be based around a different set of categories centring on the notions of 'respectability' and 'responsibility':[2]

---

2   This is supported by research by Corbett and Grayson (2010, 6) who also found that drivers described themselves as 'responsible' and 'normal'.

> I am a respectable upstanding member of this community and I object to being treated like a criminal. (Female, late 50s, Speed Awareness Course)

> I have never been in trouble with the police in my life but now suddenly I'm a criminal? I am not. I am a hard-working, respectable citizen and a very responsible person. (Female driver, mid 50s, convicted driver focus group)

These notions of respectability and responsibility are unpicked further below, but emerge as variously centring around a position within the 'moral', 'law-abiding' 'majority' of 'decent' and 'responsible' citizens, who are both 'employed' and 'tax-paying'. Such a respectable identity is, however, based on assessments of position that predate a concern with risk as a justifying rationale for enforcement. This is indicated by the repeated use of the past tense in the following quotation:

> I honestly think, speaking honestly, that until now I'd never feared prosecution. Seriously, because of what I do. I don't engage in criminal behaviour. I'd always felt, I suppose, untouchable because I'd never thought the police would have any interest in me as a respectable bloke, you know? I don't willingly behave in a criminal way, and because of my awareness of what's right and wrong and not needing to do anything wrong to improve my life, I honestly wouldn't ever have thought about it. (Male driver, late 20s, experienced driver focus group)

The shift to enforcement priorities determined by risk, however, has meant that such individuals are no longer insulated from police attention by their respectability or attitude to the law. The 'suddenness' that the second female driver, above, refers to demonstrates their shock at the irrelevance of this kind of respectable and law-abiding self-identity to new enforcement priorities.

The following analysis considers how current speed limit enforcement policies are experienced as challenging to such previously stable and secure identities. The respectable identity is divided here into two elements: those that relate to the *psychological* and those that related to the *physical* consequences of being problematised by risk-based regulatory systems.

Psychological Consequences of Speed Limit Enforcement

It has been suggested that an individual's 'protective cocoon' relies on the construction and maintenance of a reasonably coherent 'narrative of self-identity' (Giddens 1991). A concerted effort to render activities such as speeding punishable and morally condemnable seemingly challenges the individual's ability to maintain such a coherent and consistent sense of who they are, how they fit into the world and to maintain 'a tough, trustworthy, unyielding external frame' (Bauman 1997, 20) within which to operate. These challenges are considered below in the form of threats to specific elements of the respectable identity possessed and deployed by many drivers. Neither membership of 'the majority', 'morality' nor a stated intention to be 'law-abiding' offer any guarantee of a life free from problematisation by the authorities when a concern for the control and elimination of risk dictates enforcement priorities.

The 'Law-Abiding' Identity

A key element of the respectable identity claimed by some drivers was the notion of being 'law-abiding'. In claiming such a status, individuals are indicating a commitment to acting in a legal fashion and stating an intention to obey the law in future. They are also, furthermore, saying something about the way in which they would expect and believe their life should turn out, assuming that it will be free from censure and criminalisation. Individuals who describe themselves as law-abiding assume that laws will 'provide the necessary architecture in which people can plan and carry out good-faith social cooperation' (Luban 2002, 296), which will entitle them to live freely and without interference from legal authorities.

The use of risk as the justifying logic for enforcement, however, produces a situation in which a stated intention to be law-abiding is no longer sufficient to protect an individual from being criminally problematised. Behaviour is rendered punishable if it is seen as risky *regardless of whether or not it was intentional*, and the use of strict liability laws means that 'law-abiding' individuals who never intended to break laws can nonetheless be punished as law breakers. Because of this apparent challenge to the law-abiding self-image so valued by some drivers, many were quick to differentiate between *types* of law breaking. 'Crime' was therefore reconceptualised as offences for which *mens rea* was necessary, allowing a law-abiding identity to be preserved intact despite evidence of law breaking. 'Real' crimes could therefore be redefined as those for which conscious intent was required, with *mala prohibita* offences such as exceeding the speed limit reclassified and thus rendered consistent with a law-abiding self. The following exchange between two drivers evidences this important distinction:

> *Steve*: I mean, we've all broken a law, of a kind, that's why we're here, but we're not career criminals.
>
> *Annie*: But if we keep breaking the law, we will be!
>
> *Steve*: No, I mean *real* criminals who have broken *real* laws. Not accidentally gone at 37 [mph] in a 30 [mph limit].
>
> *Annie*: A law is a law.
>
> *Steve*: No, there are some laws ... there are differences. Different people commit them. (Speed Awareness Course)

Steve's approach to the law means that he is able to maintain a distinction between the type of 'law-abiding offenders' created by risk assessments and strict liability legal practices and genuinely intentionally criminal individuals. The latter group is considered to lack the inclination to be law-abiding and as such is reassuringly different to those who live their lives according to this stance. 'Real' laws, it would seem, are those that identify 'real' criminals; they do not catch out people who consider themselves to be law-abiding as a cornerstone of their identity.

For some people, however, this law-abiding identity is such that it seemingly overrides objections to the law. It can therefore be used to explain both vehement opposition to and careful compliance with the same policy. For the following compliant individual, her desire to maintain her status as inherently law-abiding meant that she obeyed laws, whatever form they took. Although her stance may be viewed as diametrically opposed to the other, disgruntled, drivers discussed here, in that she refuses to break speed limits and has her behaviour effectively dictated by the enforcing authorities, this non-offending driver can still be seen to be motivated by elements of her identity – elements that centre around her basic attitude to the law. *Her* preconception of who she is and what she does in respect of laws dictates that she does not break the law *because it is the law and she is not a law breaker*:

> *Claire*: It upset me, getting caught, because I am a law-abiding person. I don't choose which laws I think apply to me, I am just the sort of person who sticks to the rules.

> *Rob*: I wouldn't beat yourself up about it. It's not like you've mugged someone. Technically, yes, you broke the law, but you weren't arrested or anything. (Speed Awareness Course)

Claire has found herself in a confusing and somewhat distressing situation given that she is committed to law-abiding behaviour but has nonetheless found herself on the wrong side of the law. Her commitment to law-abiding behaviour has not protected her from accidentally offending against a strict liability law. Rob's approach is more matter-of-fact in that he, like Steve above, maintains that there are different kinds of law. Being on a course for convicted offenders is less unsettling to him because he maintains that speeding laws are not 'proper' laws.

Although essentially conforming individuals, like Claire, are committed to respond to speed enforcement in exactly the opposite way to their critical counterparts, the underlying reason for their stance is still rooted in their notions of self and their identity as law-abiding. Significantly, their support is still *not* explained by their acceptance of the authorities' suggested causal linkages between speed and risk but by their belief in the authorities' legitimate right to dictate behaviour. Both reactions to enforcement are explained by the desire to maintain a coherent sense of identity in the face of challenges to it posed by speed limit enforcement justified on the grounds of risk.

The 'Moral' Identity

O'Malley has observed that, in cases such as speed limit infringements, 'where the regulatory fine is applied, the generalized moral character and liberty of the offender is not the issue. Rather, it is the continued licence to engage in certain practices' (2009a, 83). However, while moral character may not indeed be 'the issue' for enforcers, meaning that it is not relevant to the successful prosecution and punishment of speeders, it appears that it remains very much the issue for some drivers.

A second significant component of the respectable identity adopted by some drivers was the sense that they were moral citizens who made moral choices. As such, they assumed that their stance insulated them from accusations of immorality. Enforcement justified on the basis of risk, however, also presents a challenge to this assumption. Regulations that are justified on the basis of risk assessments are, nonetheless, *moral* judgements given that they relate to judgements about desirable and undesirable outcomes, and contain an image of 'how we want to live' (Beck 1992, 58): 'Assessment of the chance of adverse consequences also depends on morality. Identification of a threat or danger, and of adverse consequences, is based on judgements about "goodness" and "badness" and distinctions between right and wrong' (Ericson and Doyle 2003, 2).

The increasing regulation of road use based around risk premises has therefore also seen its increasing moralisation. Actions alleged to result in risk are reconceptualised as 'bad' rather than unfortunate, leading to reductions in the use of the word 'accident' with its implications of bad luck and unhappy chance:

> [W]ith the rise of risk discourse, 'accident' is being replaced with other terms that allow the attribution of responsibility. For example, vehicle insurers substitute 'collision' or 'crash'. This new moral language is accompanied by claims that over 90 per cent of all 'crashes' result from individual drivers' actions. Road conditions are dismissed, as is the view that accidents 'just happen'. (Ibid., 7–8)

That this change in philosophy has occurred in relation to speed enforcement is clear in the way the following Chief Constable describes the thinking behind the NSCP:

> We are standing up and saying, 'It isn't just "one of those things", this is human error, this is fault-based. You have not made a mistake, you've more or less done it deliberately, or recklessly, and you've endangered the life of someone else. And that's the criminal law, pal, so the cops are going to get involved.' For the first time ever I've managed to get the DfT to talk about road *crashes* in their most recent publication, not *accidents*. It doesn't make sense to call them accidents if we are to allocate blame, because this is not just about finding causes, this is about finding *who's to blame*. (Interview with Chief Constable 1)

Reinforcing this sense of morality is a series of educational campaigns, run alongside the camera-based enforcement policy, which emphasise the driver's moral obligations to his or her passengers and other road users. The driver is invited to choose between speeds that are (often) portrayed as killing or injuring a child, and speeds that (it is suggested) will not.[3] Choosing to speed is thus presented as an immoral choice taken despite the outcome shown to result. *Mala prohibita* offences are now couched in terms that imply that they are innately *wrong things to do*. To be assigned the identity 'risky' is therefore also to experience a *moral* criticism – something that inevitably impacts on the individual's already unsettled sense of identity.

Moral overtones are especially strong where the chosen method of enforcement is, as in this case, the criminal law. Lea notes that 'criminalization' should be seen as more than 'simply a tactic for dealing with groups of individuals who constitute obstructions' (Lea 2002, 139). The police, so often turned to as the enforcing agent in questions of risk (Ericson and Haggerty 1997), can be seen as carrying with them a symbolic meaning or 'aura' that renders the issues they touch upon moralised (Loader and Mulcahy 2003, 33). The moralisation of traffic offending within a strict liability framework means that many people will be morally judged for offences that did not in themselves represent bad moral choices. Given that offending can be unintended *and* result from an intention to act morally, a moral stance is no longer sufficient to insulate a driver from moral criticism, nor from criminalisation.

Membership of the 'Majority'

Hunt describes the process whereby risks come to be regulated in the following way: 'Deviance or transgression is constructed as "risky" and becomes moralized as a possible prelude to instituting some regulatory intervention' (2003, 178). The above analysis has noted the importance of assumptions of immorality and transgression in bringing about challenges to respectable identities. In a final sense, however, Hunt's description also makes clear the assumption of a connection between risk-based control and *deviant* actions. He assumes that behaviour that causes an increased risk of harm, in whatever context, is that which deviates from the norm: that risky behaviour is deviant behaviour, and that safer behaviour is that engaged in by the majority. Majority status is, however, claimed as a significant component of the respectable identity possessed by some drivers and deployed in response to accusations of riskiness.

Research shows that speeding is a *majority* activity (Corbett 2003, 111; Stradling et al. 2003; Musselwhite et al. 2010). This presents an interesting context for regulatory interventions usually deployed, as Hunt suggests, against a minority

---

3   The 2005 campaign 'It's 30 for a reason' featured advertising that focused on the differential outcome of hitting an eight-year-old girl at 40 and then 30 mph (DfT 2005d), while its predecessor featured the slogan 'Kill your speed, not a child' (DfT 2005e).

of both morally and statistically perceived *deviant* people. As such, risk thinking challenges the elements of identity based around being part of a respectable, law-abiding *majority* as these elements of the identity are direct contradictions in the case of speed enforcement. The statistical norm, according to speed limit regulation, is deviant.

Criminal, deviant or immoral people can, in the speed enforcement context, no longer be used as something to define ourselves against, as 'collective efforts to define cultural borders and a sense of place and security within them' (Ericson and Doyle 2003, 14). Through the enforcement of traffic laws based around a concern to reduce risk, the role of the police as defenders of a majority of respectable or law-abiding citizens from a minority of the criminally deviant is therefore called into question. There is no opportunity for speeders to function as 'othered' folk devils or to serve as a group 'against which mainstream society could reaffirm its own solidaristic identity' (Sullivan 2001, 30) as it is the mainstream society itself that is being accused. The police, always seemingly protecting the majority from a minority of deviants, 'from the barbarian within', 'the enemy', 'the bad guys' (Kleinig 1996, 24), are now experienced by many in the role of *offender*. Adopting a 'risk' approach has resulted in previously moral, majority and law-abiding populations being drawn in to the systems of law enforcement. Such identities, which used to be sufficient to deflect or resist responsibility for criminal acts, offer no such protection from a concern to reduce risk.

Physical Consequences of Speed Limit Enforcement

The respectable identity introduced earlier also includes a commitment to day-to-day physical behaviour that demonstrates a respectable approach to life. Respectability is therefore not just about a mental stance but also about a set of actions. It requires physical mobility, the ability to engage in employment and the act of paying taxes. It is these actions and the way in which speed limit enforcement challenges their pursuit that form the content of the next sections. Again, the risk context is seen to magnify these consequences, and enforcement is seen to impact upon the sense of respectability of those it deems problematic.

Threats to the Driving Licence as a Symbolic Security

The ability of speed limit enforcement by speed camera to constitute a *physical* threat through potentially leading to a driving ban will be considered here. Such a ban can result from the accumulation of 12 or more penalty points on the driving licence within three years, and concerns about points generally preceded, and in some cases displaced entirely, any mention of an increased risk of a collision associated with speeding:

Sue: It's the 'points', undoubtedly. The money I can always come up with, but the points have a real impact.

Chris: I know what you mean. That's the bit that really scares me. Four times and you're banned. No questions. (Convicted driver focus group)

The money is a pain but it's not half as serious as the points that you get. The fine is like a tax which you pay and get on with it. But I think what most people are afraid of is getting points and being banned. (Female driver, late 30s, experienced driver focus group)

Chris: Twelve points always used to be the limit when it was almost impossible to get stopped that many times by a policeman, but now you can do it in one journey and that's it – banned.

Laura: All my points have come on my journey from work to home and there's *30* cameras on that journey, so as far as I'm concerned I have more chance than anybody!

Terry: You've done well, then!

Laura: It's the fact that there are so many, I'm *going* to get caught, and eventually – well – banned, I suppose. (Convicted driver focus group)

A driving ban was viewed as a serious consequence of speeding because mobility (as and when required) was seen as necessary for day-to-day physical freedom. Such freedom had obtained a symbolic importance and was seen to reflect on the 'worth' of the individual:

I think for me the big worry about getting banned is the sheer embarrassment of having to *beg* for lifts off of people: friends, family, all of them. Having to rely on them to carry me around, when it suits them, and feeling like I constantly owe them something. (Male driver, early 20s, Speed Awareness Course)

[Losing my licence] would make me a burden on my friends and family. I couldn't pull my weight with the kids any more. The wife would have to do it all, shopping, visiting, taking them to school and all that. I'd be no use at all. (Male, mid 40s, experienced driver focus group)

This was even the case for young drivers, relatively new to the experience of being able to drive but already dependent on the freedom it gave them:

My big fear is having to use public transport which I quite honestly find dirty and disgusting. I'm sorry, I know that sounds terrible but it's true. The whole idea of having to wait for a bus, then stand up, or worse still sit next to some oddball, having to turn up at my friend's house having got off the bus. Yuk. I mean, what sort of people have to get public transport these days? (Female, late teens, new driver focus group)

Flexible, independent mobility was seen as necessary for a respectable existence, with those without the symbolic security of the driving licence viewed as somehow less useful, less civilised, less 'normal'. Speed limit enforcement therefore threatened what, to many people, was a basic requirement of a functioning human existence. Corporeal transport is, as Urry suggests, 'essential for constituting social and economic life and ... not an optional add-on' (2002, 263) and also has a symbolic importance: '[risk society] regulates the pace of contributions to society by rating credentials, personal handicaps, creditworthiness, productivity and so on. Some people are destined for the autobahn, others relegated to highways with speed limits, and still others to local roads with speed bumps' (Ericson and Haggerty 1997, 40). To extend this (extraordinarily appropriate) metaphor, therefore, the ultimate in exclusion is represented by the denial, revocation or suspension of the driving licence, denying access to the roads in any form. The driving licence is seen as one of risk society's near-essential credentials, which testifies to an individual's abilities and which allows them to participate as a full and active member of society. This is despite the fact that, as O'Malley notes, the ban is only intended to target the individual – to ensure that the dangerous are 'removed from circulation by having the specific risk-producing capacity neutralized' (2010, 803).

Employment and the Viable Human Resource

The threat of exclusion as a result of licence loss was also understood in terms of the threat to employment status that repeated prosecution under the NSCP could bring about. Such employment was a further necessary component of a respectable identity, and its loss was seen to impact negatively on the individual's 'viability' as a 'human resource' (Ericson and Haggerty 1997, 197). Recent and ongoing changes in the nature of employment have led to a requirement that all employees are both flexible and mobile, able to respond to changes in demands placed upon them, or otherwise replaced by other more competitive (more flexible, more mobile) individuals (Beck 1992, 94). Failure to adapt is, furthermore, according to Beck, transformed into a personal failing, reflecting badly on the individual who could not keep up, rather than on the system that made him or her expendable (ibid., 89). Mythen notes that changes in the experience of employment have been singled out 'as a decisive factor in the development of uncertain and insecure forms of lived experience' (2005, 130), with 'taken for granted assumptions' about the individual's place in society being undermined by the prospect of being without employment (ibid., 133). The increased attention being paid to speeding offences means that the likelihood of being banned from driving is increasing at the same time as these changes in employment markets are also occurring. As such, many drivers felt that the ability to obtain and retain paid employment was increasingly dependent on the possession of the driving licence and the flexibility, reliability and competitiveness that it made possible:

> The money doesn't bother me really – I thought 'fair cop', I was speeding – but it's getting banned; to get banned is a bit rich. There are people who will find it a heck of a lot more difficult [to find employment] and all they've done is go a little bit over the speed limit. Why should they have the worry of it? They've done nothing wrong really and they're being threatened with their livelihood being taken away from them! (Male mid 50s, convicted driver focus group)

> The first thing that came to my mind was the 'licence' situation! Losing my licence and not being able to get to work, so losing my job. Especially when I was on nine points, I could lose it at any time and that would be that. Then it would be the accidents, or the chance of getting injured, but the first thing would definitely be my licence. (Female, late 20s, convicted driver focus group)

> For me it's the loss of licence, yeah, loss of licence, because I need my licence to work basically. I mean, if I didn't have my licence there would be ways round it but I don't know. Presumably I'd be reported to the GMC [General Medical Council] as well, I imagine, for losing my licence. But there again you've only got to fart and you get reported to the GMC these days. Any excuse. It's ridiculous, but, anyway, the licence is the first thing. (Male, late 40s, convicted driver focus group)

Although Mythen criticises Beck for suggesting that this process of destabilised employment is universally experienced, suggesting that class is still a key determining factor in the extent and effect of this instability (2005), none of the individuals who mentioned employment risks did so in an overtly 'class'-based manner. Instead, the near universality of the need to be mobile and flexible in order to be a viable and productive member of society is assumed. The importance of the possession of legitimate employment was, for example, noted in focus groups by an individual employed in retail, a shift worker and (as in the final example, above) a medical professional – none of whom had explicitly driving-related forms of employment.

As such, the loss of the driving licence is clearly the dominant concern, set here within a developing sense of the increasing vulnerability of employment – even for professionally qualified persons. Although financial penalties impact differentially on different income groups, with the better-off potentially able to absorb such fines as merely inconvenient, each driver has a limit of 12 points and is equally susceptible to a driving ban. Speed enforcement is therefore one way in which all classes, trades and professions are rendered equally vulnerable to criminalisation and its consequences, because, quite simply, they are all deemed to pose the same amount of risk in exceeding the speed limit.

Exclusion via unemployment was a fear for many drivers who needed to commute to work and felt that their car was essential for this, but it was particularly acutely felt by professional drivers. Such drivers are defined here as those whose driving was an integral part of their role and for whom their licence represents their actual credentials and qualifications for their job itself. Such drivers were in no doubt that their employers were interested only in their sanctioned and licensed ability to perform their tasks, and would show no loyalty if one of their staff were suddenly rendered 'useless' in this respect as a result of the proliferation of camera enforcement. The professional drivers quoted below expressed concerns that can be seen to confirm Mythen's view that 'responsibility for employment risk [is] being transported away from employers and toward employees' (2005, 143):

> *Dave*: It doesn't affect the company when you get points tallied up and lose your licence, but it affects you personally, doesn't it? Mental worry *all* the time.

> *Pete*: I don't think they care, really. At our company they check our licences every six months, so on that basis they want to check if you're legal. I don't suppose they really care, really. If you're banned, you're banned. You're a number.

> *Ray*: They just bring somebody else along. That's all they do. (Professional driver focus group)

> We've had in the past, we've not employed anybody, but we've certainly had applicants who've been disqualified for different reasons, but our broker will want to put a three or four hundred pound additional premium on them. It's a discouragement to employ somebody. (Male, mid 50s, professional driver focus group)

This was despite the fact that many employers were considered to be complicit in the criminalisation of their employees, setting targets and arranging schedules that were achievable only by breaking speed limits and then dismissing drivers who were caught:

> *Mike*: We do have an increasing problem with speed cameras and speed humps and the like because we are under pressure to meet targets, and it's difficult to get the company to reflect the target with the cameras and all that.

> *Alan*: For an express van driver, they're under pressure and there's never any reflection in the targets to reflect the fact. We're not against speed cameras, but the targets that we have to do in a day should reflect this now, and yet there's the same pressure that there was years ago before all these restrictions.

Pete: I have asked them [employers] to change it, but they come back with the same 'Look, the average driver gets back in nine hours, your targets are right'. But there's no reflection of how he's tearing around to do that! Or what accidents he's had or how many speeding fines he's getting. They're not interested in that, just meeting our targets. (Professional driver focus group)

For all drivers who felt that their continued or future employability was related to their ability to drive, the associated costs of losing that employment were high and were an ever-present concern:

If I get any more points I lose my licence and as a result will lose my job. Then I can't pay the mortgage so my family will lose their home as well. I can't see my wife putting up with that! It's as simple as that! (Male, late 50s, professional driver focus group)

They [points] take on this added importance. It focuses your mind because if you get done for speeding four times, 12 points, that's it: no licence, no job, no way to pay the bills, even if only for a short period of time, but enough to be giving you issues about getting reinsured and restrictions on you driving vehicles in the future. (Male, early 40s, professional driver focus group)

For many, this threat was viewed in terms of their increased exposure to cameras, something that made them more likely to be detected speeding on occasion. The chances of losing the licence were therefore multiplied by the nature of the job, something which, in turn, multiplied the consequences of that loss:

The way I see it is that I'm out there more, doing more miles, passing more cameras, so the odds are against me if you like. Chances are it'll be me more often than it is you. But for me it means my job. (Male, late 30s, Speed Awareness Course)

As these professional drivers experienced it, this policy was seemingly trying to turn them from incorporated, included citizens into incapacitated welfare-dependants, to brand them with 'the stigma of the impotent and the improvident' (Bauman 1997, 37). The apparent remoteness of the risk of an accident was supplanted by the very real and immediate threat represented by losing the driving licence. Each set of points represents an increased vulnerability, another step nearer the edge, of 'tipping' into exclusion. Speed cameras therefore become 'recurrent switch points to be passed in order to access the benefits of liberty' (Rose 2000, 190). The more cameras there are, the more it becomes likely that the individual's behaviour will fall below the required standard at one of these switch points, rendering the maintenance of full access more and more difficult at a time when it is of increasing importance. The result is that 'any individual is only one technological innovation away from losing his or her privileged status' (Baker

2003, 275). As such, even 'those "lucky" enough to be "included" are not part of the culture of contentment'; instead, 'they are unsure about their good fortune, unclear about their identity, uncertain about their position on the included side of the line' (Young 2003, 399). The role of the driving licence in generating this insecurity is, then, as a tangible permit, serving as a physical manifestation of the more abstract threat to the identity engendered by the use of the criminal law as the regulatory method.

Total, independent mobility can therefore be seen a necessary, though indirect, prerequisite for wage earning and in turn for participation in consumer markets, for purchasing goods and services, for 'active citizenship' (Rose 2000, 190). This notion of active citizenship, however, does not just involve *purchasing* goods and services. It also involves an awareness of the contribution that that citizen makes to society in the form of taxes, as the following driver demonstrates: 'Well, if they ban me, then I can't work and they can't have my taxes. If they do that to all of us, then there'll be trouble' (male professional driver, late 40s, Speed Awareness Course). Paying taxes may also give the contributing individual a sense that they should have some control over the way in which those taxes are spent. It is unlikely that this bargain includes the authorities using those taxes to criminalise the same individuals who supplied them with funds and, as is often claimed, paid their wages. The paying of taxes that fund the police is, seemingly, part of the deal that 'should' insulate drivers from becoming part of the police's enforcement focus. The sense of 'ownership' to which Girling et al. refer, below, should perhaps not therefore be understood solely in a protective sense. It also provides the opportunity to exercise control: '[T]raffic policing is unsettling because it offends against many (middle class) people's sense of themselves as the proper recipients of police services, and jars unpleasantly with the sentiments of attachment and ownership they feel towards the police' (2000, 137). This view of the police as 'belong[ing] to "*us*", to be directed at "*them*"' (ibid., 137) is threatened by policies that shift the focus of attention on to those not used to it. Communities more used to treating the police as a resource and as a feature only present when *they* summon them, and which otherwise leaves them to live their lives without intervention, may, as a result, experience dissatisfaction with their experiences of either *under-* or *over*-policing. It is just such concerns that we turn to below.

The Offender as Victim

The challenges posed by speed limit enforcement to the two aspects of respectability discussed above have produced a situation in which some drivers are able to view themselves not as *instigators* of risk but as *victims* of it: 'I just don't feel in the least bit naughty about it. I feel I've just been, I'm just a victim of the system' (banned male driver, mid 50s, convicted driver focus group). Such a sense of victimisation is made possible in two ways: first, because drivers have something that they feel is put at risk by the policy, and, second, because this takes place in a context in which there is dispute about the legitimacy of that regulation in the first place. The policy thus brings about harmful consequences for individuals which many are not convinced are justified. Regulation is therefore able to be experienced as of questionable validity and can come to be viewed as victimisation rather than protection from risk. Some drivers have thus been able to reconceptualise themselves as more 'at risk' of *speed enforcement* than of *speed itself.*

That this inversion is not only possible but also a potentially attractive alternative to being cast as an instigator of risk may in part be explained by the way in which such offenders are treated:

> 'Justice' is now very much less important than 'risk' as a preoccupation of criminal justice/law and order policy; the politics of safety have overwhelmed attachment to justice in the institutions of late-modern democratic polities. If someone, or some category of persons, is categorized as a risk to public safety, there seems to remain scarcely any sense that they are nonetheless owed justice. (Hudson 2001, 144)

Defending oneself against an accusation of riskiness is difficult, if not impossible, in such circumstances. The adoption of the role of injured party, by contrast, offers the individual opportunities for self-defence. In fact, such action has been positively encouraged by the state. Individuals have been required to become active in securing their own protection from risks, knowledgeable about their own circumstances and prepared to act to self-govern in relation to them:

> Everyday risks present us with the necessity of making a seemingly never-ending set of choices. The significance of these choices is compounded by the disparate pressures of the mechanisms of responsibilization that demand that we make them in a context that requires us to treat our lives as a project over which we should exercise a deliberate and long-term calculative effort. (Hunt 2003, 169)

Such 'responsibilization' is understood here in terms of *the individual protecting themselves from external risk sources*.[4] Individuals are 'encouraged, provoked, and incited to engage in taking care of themselves' (ibid., 181), to the extent that '[i]n late modernity *not* to engage in risk avoidance constitutes a failure to take care of the self' (ibid., 182). As a result of this process, anyone who is, for example, involved in a collision is rendered responsible for it; he or she is an individual who 'failed to take adequate preventative measures by learning and acting upon knowledge of risk' (Ericson and Doyle 2003, 8). What this also reinforces, however, is the view

---

4   By 'external' I mean originating from outside of the individual and impacting upon them, not 'external' in Giddens' sense of the more naturally occurring risks which predated manufactured risks (1999, 4).

that the individual is primarily an entity that is *at risk* rather than one that *poses* risk to other entities. This is epitomised by Garland's description of what risk *is*: 'The risks we run depend on the actions of others and the risks they take. Steering clear of dangerous drivers is the simplest example' (2003, 55).

Clearly risks, in this explanation, are dangers to which 'we' are exposed to by 'them', and the following description of the requirements of responsibilisation placed on the individual again supports the imperative of taking responsibility for risks that threaten us and our families. It is, however, silent on the issue of taking responsibility for the harmful acts that we engage in ourselves and which constitute risks to others:

> As far as individuals are concerned, one sees a revitalization of the demand that each person should be obliged to be prudent, responsible for their own destinies, actively calculating about their futures and providing for their own security and that of their families. (Rose 2000, 186)

In the traffic context, however, this view of the essential externality of risks is particularly hard to sustain. The individual is simultaneously *at risk of* others' behaviour and *posing* a risk to those same others in return. In response to this new and unsettling situation, drivers will be shown to use various strategies that allow them once again to externalise the sources of risk and to decline to accept their role as instigator of it: to cast themselves in the role of potential victim once more. *Responsibility* is declined, while methods of *responsibilisation* are enthusiastically adopted.

Responsibilisation via Deresponsibilisation

One approach to responsibilisation in the face of risks (however perceived) is to resist being made responsible for them in the first place. Such an approach is seemingly encouraged by the emphasis on risks to which we are exposed at the expense of risks that we pose to others. Hunt calls this shifting of blame *de*-responsibilisation, noting that 'people both seek responsibilities and just as strenuously refuse to accept responsibilities, processes that are the result of viewing life through the lens of risk discourses' (2003, 186). Such deresponsibilisation is, therefore, a responsibilisation strategy in itself given that the shifting of blame for risk is one way of avoiding being made responsible for risk. It therefore protects the individual from the harmful consequences – or 'risks' – of being identified as risky.

Attempts at obtaining protection from the negative consequences of being implicated as 'a risk' are, in one sense, made via a strategy that emphasises the risk-producing behaviour of others in an attempt to de-emphasise the risk posed by speeding drivers. As 'they strike out blindly at anything that gives off the scent of deviationism' (Beck 1992, 12), drivers can then relocate their own behaviour lower down a list of problematic behaviours, less 'bad' and less 'risky' than that engaged in by others. Through this strategy, some drivers are able to revert to the role of victim through the implication that respectable citizens are put at an increased risk of victimisation from traditional criminals. This increased risk is brought about by the authorities' perceived neglect of these more 'worthy' enforcement targets in favour of motoring offenders. As such, this victimisation is brought about in a direct sense by the perception that such 'real criminals' are allowed to continue to offend unhindered by the police, but also in an indirect sense whereby the authorities are considered to be complicit in this. The authorities' distraction from their 'proper objects' of control (Fiske, quoted in Coleman 2004, 8) means that they contribute to the victimisation that those proper objects can thus bring about. The goal of the strategy is to force a rethinking of police priorities that reinstates the implicated instigator into the role of potential victim, *protected by* and not *at risk of* the actions of law enforcement agencies. By attempting to influence the enforcers' priorities in another direction, drivers are therefore intending to bring about their own protection from the risk of speed limit enforcement. Effectively, therefore, those employing this tactic can be seen to have become *so* responsibilised that they use this tactic to protect themselves from external risks to their respectability. Deresponsibilisation is therefore a responsibilisation strategy that protects individuals from being labelled as risky.

It is crucial to note that this demand for the police to prioritise other offences is understood *within the context of speed enforcement*. By this I mean that these protests are not simply to be understood as the familiar call for 'more bobbies on the beat' to offer the law-abiding majority protection from burglars, for example. Instead they are demands for a reordering of priorities in which the (known to be limited) resources of the police are redeployed on to these more traditional targets and away from those who consider themselves to be respectable. As a contrast with the (often) unintentional, majority activities of speeders, two forms of intentional, (statistically) deviant criminality are made the focus of this strategy: 'real criminals' and other, differently understood, road-user behaviours.

'Why aren't they out catching burglars?': Focusing on 'real' criminals The driver's role in causing risk is, through this deresponsibilising responsibilisation strategy, relegated in importance by juxtaposing it with worse criminal acts. This approach helps to reinstate potential speeding offenders in the more comfortable and less identity-threatening position of potential victim of an external threat, by situating their own offending within the context of 'real' crimes to which they fear becoming a victim. As Gabor suggests, the 'victim' role (that adopted in relation to 'real' crime) is more reassuring than reflecting on our own activities:

> [W]hereas concentrating on the most repugnant acts enables us to view criminals as the personification of evil or as psychologically maladjusted, focusing on more mundane forms of criminal behaviour entails looking more closely at ourselves. Stereotyping criminals as murderers, rapists, and muggers distances

us from the criminal and sets up a simplistic view of the world as one in which people are either decent citizens or villains. (1994, 4)

This deresponsibilisation strategy involves suggesting alternative behaviours that are seen as producing more harm and thus being more worthy of enforcement attention. Such 'proper' targets are all, crucially, offences that are perceived to be immoral and require intent, and which are statistically deviant:

> I just wish they [the authorities] would get a bit of perspective, you know? Speeding on an empty street versus knifings, old ladies getting burgled and raped, and all sorts. I know which I wish the police would concentrate on. (Female driver, mid 20s, new driver focus group)

> And what's happening while the police are obsessing about speed? Violent crime is on the up, terrorism, burglaries and I don't know what else. (Female driver, late 50s, experienced driver focus group)

This strategy was acknowledged during Speed Awareness Courses, with the convener asking: 'When you heard about this course, did you all think, "Leave me alone and go and catch the real criminals. Aren't there enough burglars in Staffordshire?"' which was met with a unanimous 'Yes!' This was followed by a deliberate exercise in distancing the course and the course leaders from the police in order to challenge the perception that speed limit enforcement undermines the fight against 'real' crime.

Research by Blincoe et al. also found that speeders' responses often took this form, with demands that the police refocus their attention back on to 'real criminals' (2006, 374). Such responses to speed limit enforcement suggest that these accused drivers not only sense that traditional criminality is left to grow unhindered, but feel more keenly the loss of the police as a protective agency, something exacerbated by their new role as aggressors. There remains, however, a stubborn belief in the police as *the* solution to social problems, which makes their refocused attention on to their previous supporters all the more unsettling. They are simultaneously a reminder of the threat posed by traditional criminals and a visible indicator of the changed priorities underpinning enforcement in risk society. Furthermore, they have lost the reassuring and protective associations identified below as acting as a counterbalance to images of vulnerability that they conjure up:

> To speak of policing is, on the one hand, to conjure a reminder of the existence of the undesirable, criminal Other. As such, it is apt to prompt, or at the very least reinforce, feelings of hostility, aggression, powerlessness, violation, and vulnerability; a sense that one's security is contingent, the social world a rather fragile and troubling place. Yet the idea of policing also brings to the fore sensations of order, authority, and protection; it makes it possible for people

to believe that a powerful force for good stands between them and an anarchic world, that the state is willing and able to defend its citizens. (Loader and Mulcahy 2003, 43–4)

The increased sense of vulnerability from the traditional criminal, magnified by the absence of the usual protection, is set within the context of an enduring sense of general malaise that has supposedly beset Britain and has made its citizens feel ever more insecure: 'I feel like at least I've done my bit for the budget deficit. They can have my £60 and use it to take illegals on holiday or whatever they want. I gave up on this country a long time ago' (male, 50s, Speed Awareness Course). The 'illegal' (immigrant) and the criminal are viewed as less worthy recipients of favourable treatment – treatment that is denied to the law-abiding, respectable majority who feel unfairly targeted by speed cameras. The contrasting fortunes of the two groups are volunteered as evidence of the changed priorities that are now felt to motivate the authorities and which disadvantage those that feel entitled to a better service. The perceived neglect of such 'proper' targets as burglary and murder means that drivers can view themselves as simultaneously under-protected and over-policed, and it is their attempts to change this situation by trying to influence enforcement priorities that underpin the objections cited here.

Focusing on 'dangerous' drivers   Unlike the emphasis on 'real crimes', the second approach to deresponsibilisation *accepts* that the harm of road death and injury is a viable and legitimate enforcement target. What it disputes, however, is the way in which this harm is brought about, suggesting that other road-user behaviours are responsible for the undeniable reality of road crashes. In constructing other road users as a source of this harm, respectable speeding drivers can once again adopt the role of potential victim and view themselves as under-protected by roads-policing policies that place disproportionate emphasis on the enforcement of speed limits.

Drivers keen to stress their own law-abidingness and respectability are clearly keen to point to other types of driving behaviour as being demonstrably worse, and potentially causing risk to themselves:

> Drinking and driving is a lot worse than what I did, it was only 38 mph and for me it was just a lapse. And what about these people I've seen reading/shaving/ on the phone while driving? These are the people that want stopping: they're much more dangerous than me doing 37 [*sic*] in a 30 [mph limit]. (Male, early 60s, Speed Awareness Course)

> And where's the queue always on the motorway? In the outside lane with everyone trying to go faster than everyone else. That's when the accidents happen. I like the cameras that are monitoring, not just for speeding but for bad driving, they're on the telly at night and they show the bad driving practices. I think the police are checking out that and that's good, they should be checking

out down there for bad driving rather than just speeding, people bad driving. I
think if they bring the bad driving in as well as the speeding, then road safety
would be a lot better. (Male, mid 50s, professional driver focus group)

Again, Blincoe et al.'s respondents also bemoaned the perceived neglect of
'road rage drivers' and 'the lunactics on the roads' and the over-emphasis on the
transgressions of 'the average motorist' (2006, 374). This approach is also used
by those claiming to represent motorists, such as the member of a road safety
pressure group quoted below. There are reminders of the strategies of promotion
(specifically the attempts to homogenise the audience) used in the debate, and
discussed earlier, in this statement that: 'Treating responsible drivers like naughty
children while ignoring the truly dangerous ones, is not something that any
civilised society should be comfortable with' (member of a road safety pressure
group, quoted in *Daily Express* 2010a).

A variety of alternative behaviours can therefore be put forward as more
deserving of police attention. 'Dangerous' driving practices, whether specifically
identified or not, are seen to be qualitatively different to speeding offences. They
are, by definition, 'dangerous' (causing tangible danger), not simply 'risky'
(containing the potential for harm). The alternative worthy target behaviours
are seen as demonstrating deliberate recklessness on the part of drivers, making
them culpable risk producers and viable enforcement targets. As such, any other
deviating behaviour is advocated as a more suitable enforcement target if this
helps the accused driver to deflect responsibility and to render risks 'external'
again (Hacking 2003, 25). The alternative suggested behaviours that underpin this
deresponsibilisation are, again, seen to be ones for which intent is required and
are not therefore behaviours that can be accidentally committed by the responsible
driver. They are, furthermore, offences that are seen to be both morally and
statistically *deviant*, restoring the speeding driver to a position within the moral
majority. Drink driving therefore often features as an example of an unquestionably
legitimate enforcement target:

I've heard that convictions for drink driving have gone down since they started
using cameras. People just know that they can do whatever they like on the
roads so long as they stay under 30 [mph]. I think it's disgraceful and it makes
me scared to use the roads at night, it honestly does. (Female driver, late 30s,
experienced driver focus group)

You know a few years ago when it was drink driving that was the thing? Well,
that always seemed to make sense to me. Most drivers were scared of coming
across a drunk driver at night and they managed to stop the minority from doing
it. But this, there doesn't seem to be that logic, or that will from people. (Male
driver, mid 50s, experienced driver focus group)

The minority, intentional, deviant status of the offence of drink driving has been shown to be significant in achieving its successful criminalisation. In this case, occurrences of the offence can be effectively blamed on a subset of individuals who were different not only to the majority of *people,* but, significantly, to the majority of *drivers.* Gusfield suggests that the image of the drink driver was as a member of a lower socio-economic group. As such, his/her condemnation and subsequent criminalisation serves to reinforce rather than undermine the essential difference between criminals and non-criminals (1981, 102). Gusfield suggests that, although the middle and upper classes also consumed alcohol, the problem of drink driving was successfully shifted on to a subset of 'sick' and 'quite literally morally suspect anyway' people in the lower orders (ibid., 103). In this way, 'the social structure and the legitimacy of social hierarchy are reinforced and the possibility of degradation is dispelled' for respectable individuals (ibid., 103). As such, drinking drivers are, foremost, drinkers who happen to have got behind the wheel of a car. The 'driver' identity can be preserved as intact by this distinction, which blames the behaviour on the 'master status' of 'drinker' or otherwise 'offender' (Becker 1963/1997, 33). Gusfield then makes the following comparison with other motoring offences: 'The drinking-driver is a serious malfeasant; he is not the average citizen, the "natural man" who sometimes speeds, makes a left turn against a prohibiting sign, or otherwise acts in the foolish or perverse manner of you or me' (1981, 112). Such master-status drivers can occasionally make minor errors, but their offending actually *reinforces* their status as 'average citizen' and testifies to their 'naturalness'.

In this way, the importance of maintaining an essentialised difference between 'offenders' and 'normal people' is demonstrated. This difference is challenged by the enforcement of criminal laws in relation to majority activities such as driving, but, as the case of drink driving shows, can be maintained provided that the illegal driving activity remains something that only a minority of drivers engage in. Most drivers, furthermore, do not drive under the influence of drugs, most do not neglect to tax and insure their cars, and so these activities remain deviant even within the context of driving. Their problematic status remains consistent with a view of the police as acting in the interests of a majority against the activities of a minority. The 'turning when prohibited' and speeding examples given by Gusfield are rendered consistent with an otherwise law-abiding identity given that they are 'normal' behaviours for drivers.

The position of speeding as the least 'bad' of a set of traffic offences which are the least 'bad' of all crimes (Corbett and Simon 1991) means that there is a ready stock of other offences that can be highlighted in the attempt to point out more deserving recipients of police attention. Drivers are able to see themselves as potential victims of these other types of neglected driving behaviour and of 'real' crime because the police are perceived to be interested only in detecting speed-limit infringements. Their victimisation by such criminal others is thus caused indirectly by the authorities. Seemingly the only person deimplicated in the *causing* of risk (and yet perceiving themselves to be *exposed* to risk from all directions) is the speeding driver. Deresponsibilisation is thus very effectively achieved and accused drivers can redefine themselves as victim, vulnerable once more to externally produced risks.

Responsibilisation and Speed Enforcement

The threats to identity and mobility discussed previously also allow drivers to construct themselves as *directly* victimised by the enforcing authorities, not just by their neglect of 'real' criminals. Such individuals are able to view themselves as more 'at risk of' the actual enforcement of speed limits than they are of a speed-related crash. These individuals are also operating with a risk logic, but in this case one that rejects the causal interpretations offered by the state and instead substitutes alternative interpretations that dismiss the validity of punishments for speeding. Such drivers have not only rejected the causal interpretations put forward by the authorities, but have proposed their own risk and the causes of it. The foregoing explorations of the scenario of multiple, conflicting expert voices and of the harmful consequences made possible by being successfully labelled as a risk can both be seen to contribute to making both parts of this process more feasible and more likely. They result in a context in which *speed limit enforcement* is experienced as a more real and consequential risk than *speed.*

Responsibilisation strategies are then deployed in a very direct sense to protect individuals from victimisation caused *by the state*. Such responsibilisation has been positively encouraged in relation to traditional crime, where, as well as being knowledgeable about the risks we face, we are also increasingly required to take action to protect ourselves against them: 'With the diminished role of the State as agent of social welfare, individuals are encouraged to govern themselves through protective services and products offered for sale' (Haggerty 2003, 194). Home owners are encouraged to install burglar alarms, car owners to purchase alarms and locks, computers users to protect themselves against viruses, and pedestrians to be aware of their own personal safety as part of their 'private, defensive routines' (Garland and Sparks 2000, 16). As a result, those who fail to take these kinds of precautions can be portrayed as complicit in their own victimisation, with blame falling directly on them rather than on the authorities. The growth of the private security market suggests that this is a responsibility that individuals are accepting with some energy, and that the message that the state has 'backed off' and can no longer be solely responsible for crime control has been heard and acted upon (Garland 1996). However, the above examples all relate to situations in which the state and the individual agree as to the harm that needs to be protected against, and reinforce the 'us' and 'them' relationship between the law-abiding majority and a minority of the criminally intentioned. As has been shown, this relationship is undermined by the current speed enforcement policy, with the result that these clear distinctions are clouded and subject to debate. In this context, it is the state itself that is viewed as the risk by many individuals – individuals who have previously been encouraged to adopt a proactive role in relation to their self-preservation and protection from risk. Having provided this encouragement, the state now finds itself on the receiving end of its own logic, with individuals adopting the identity of 'victim' and with the state recast as the aggressor, offender or instigator of risk.

This section considers responsibilisation strategies that are used to render the individual less physically vulnerable to the risk of criminalisation and its consequences. Such interventions therefore directly protect the driving licence and also prevent the individual from being identified as a risk, thereby removing both the unpleasant physical and psychological consequences of enforcement. As has been shown, the consequences of this 'victimisation' are real and serious enough to make interventions that offer protection an attractive proposition. This preventative, protective action has taken at least six forms in relation to speed enforcement by speed camera. These responses are explored below, with the final strategy – that of direct action tactics against the technology – representing the most extreme example of driver 'responsibilisation'.

Strategy 1: Camera jammers and deflectors   Parallel to the proliferation of speed cameras has been the development of a range of devices that offer to prevent detection by them. Such devices effectively render their owners less vulnerable to detection for speeding by interfering with the radar beam that is used to detect offences. The market for such devices is thought to have grown at a rapid rate, although its unregulated nature makes this hard to quantify (DfT 2005a, 22).

In a different context, individuals who purchased such protection from risk would be viewed as acting responsibly, *accepting* responsibility for their own protection and for the protection of their homes, jobs and families. However, the fact that the risk to which they are responding originates with the authorities means that the purchasing of such protection becomes illegal (DfT 2005f, 22).

The availability of such technologies can be seen, however, as a logical response of the market to an emerging threat that targets many of the individuals who have similarly taken the same 'responsibilised' attitude to their homes, computers and leisure time. In terms of protection from a threat that is experienced not just as inconvenient but as personally incapacitating, the purchase of a speed camera jammer and a crook-lock are motivated by the same concerns. They differ only in the source of the perceived threat.

From the state's perspective, however, operating with the causal logic that it does, the purchasing of such technologies allows drivers to behave in a more risk-producing fashion by enabling them to speed with impunity. The prohibiting of such technologies is therefore motivated by the desire to maintain the effectiveness of its own protective technology (the speed camera) and thereby reduce risk. The legislation is itself a crime-prevention intervention in that it is designed to prevent offences of speeding taking place and targets technologies that it believes make offending more likely, resulting in secondary deviance and producing an amplification of criminal activity more generally (Fox 1995, 77).

As such, responsibilised drivers who opt to protect themselves from the risks that they experience as potentially threatening to their identity, status, homes and livelihoods are rendered even more vulnerable by taking these actions. As well as being denied the right to protect themselves, they become subject to further laws that threaten to punish them again for their 'responsibilised' actions.

Strategy 2: Camera detectors   Devices that do not interfere with speed detection equipment but notify drivers of its presence are similarly available to drivers. Such devices are, however, legal and there are no proposals to change this status (DfT 2005a, 21). Speed camera detectors are marketed in similar ways to the illegal devices discussed above but offer instead to notify drivers of an approaching speed camera so that they can modify their speeds to within legal limits. These detectors operate in a similar way to the various legal methods already in use to signpost the locations of cameras, such as official roadside signage, local newspaper announcements about enforcement venues or the inclusion of camera locations in road atlases. Research by Corbett and Grayson reports that 'a sizeable number of drivers thought it reasonable to slow down only when passing a camera or to use technical devices to warn of camera sites' (2010, 2). These technologies were found to be particularly popular with '[d]rivers who had already accumulated a number of points [who] often relied on technology to avoid getting more points, rather than simply driving within the limit' (ibid., 3). Such manipulation of the system effectively undermines the capability of speed cameras to identify and exclude from the roads those who will not be deterred, and therefore limits their contribution to risk reduction. The extent to which this is a problem for road safety depends on whether or not it is considered desirable that drivers comply with the limit at speed camera sites or do so across the entire road network. Making cameras highly visible to drivers is a similar development, brought in relatively early in the speed camera debate, which demonstrates the trade-off being sought between maximum speed limit compliance and driver acceptability.

This difference in legal status of these devices and those considered previously is explained by the fact that such devices do not permit drivers to continue to speed with impunity at speed camera locations, but encourage them to moderate their speeds at these points. Presumably, the assumption is that drivers may be speeding at other locations and thus require information as to where and when they should slow down. However, this form of responsibilisation is permitted because it enables drivers to remain within speed limits at points where this has been deemed particularly necessary and, it might be suggested, where the success of the technology will be measured.

Such products are marketed with reference to the kind of protection they offer, with the 'Road Angel' device carrying the slogan 'Protecting your life and livelihood' (Blackspot.com 2010). The manufacturer's website now warns drivers that "[s]ome 2.1 million drivers were caught by fixed safety cameras in 2005 … don't end up being the next statistic!' (ibid.). In both senses, the marketing represents an interesting blending of the two notions of risk that dominate this debate, where 'becoming a statistic' now seemingly has dual meaning. The ownership of such a device by the Duke of Edinburgh demonstrates the mainstreaming of such technologies, despite their hazy status in relation to road safety. Newspaper reporting of this fact notes that he will 'avoid trouble' by using it, again maintaining the simultaneous allusion to both crashes and prosecution (*Daily Mail* 2010a).

Strategy 3: Insurance Insurance technologies represent one of the powerful regulatory technologies that underpin the managing of potential negative outcomes in a risk society (Ericson and Haggerty 1997; Ericson and Doyle 2003; Heimer 2003). As if to further reinforce the idea that speed cameras are risks like any other to be protected against, responsibilised drivers are offered insurance against losing their driving licence. Such policies provide for alternative transport for drivers who receive short-term disqualifications for speeding, either by offering to repay the costs of taxis used during the lifetime of the ban or providing a chauffeur. As a result, although the initial detection and prosecution are not prevented, such responsibilisation strategies allow victimised motorists to reduce the negative consequences of their offending/victimisation. The multiple and serious consequences of the loss of a driving licence are emphasised, as in the following advertisement, in which it is losing the licence, not the behaviour for which that loss is the punishment, that is being equated with death: 'Must drive to stay alive? Keeps you on the road, keeps you in your job, keeps you in your lifestyle' (St Christopher 2010). The wording of advertising for such policies reinforces the perceived difference between being disqualified for speeding and being disqualified for 'dangerous driving' or driving under the influence of alcohol or drugs, with companies, for example, keen to point out that they 'do not condone stupidity or dangerous driving and that the product does not cover individuals convicted of dangerous, reckless, drink driving or driving under the influence of drugs' (ProtectionInsurance.com 2004). In an interesting twist, the product is marketed not just as 'protecting' drivers but as protecting speed cameras from vandalism, presumably by lessening the consequences of being detected speeding and thereby lessening the resentment aimed at the technology that has made it possible (ibid.).

The existence of insurance policies that serve to ameliorate the consequences of losing the driving licence helps to reinforce the impression that speed cameras are risks to be treated in the same way as any other threat that is experienced as unpleasant. To the insurance industry it is irrelevant where the risk originates if it is real in its consequences and if there is a market for its services, a fact that perhaps serves to reinforce this same belief among the industry's clients.

The role of insurance both in *penalising* drivers through increased premiums when they get points on their licences and in *helping to protect* the driver from the technology that inflicts those points demonstrates that insurable risks are anything that is experienced as such. Such risks cannot be simply handed down or prescribed and need not be unquestioningly adopted by those on whom they impact. The risk logic that motivates enforcement need not be the risk logic with which it is experienced, as the example of insurance demonstrates.

Strategy 4: Guides to avoiding fines and/or points   The purchasing of guidebooks that promise, for example, to exploit 'secret legal loopholes the police do not want you to know!' (UK Driving Secrets 2010) within the law relating to speed camera enforcement can be understood from the perspective of both the increased democratisation of expertise and the responsibilisation of the general public. Haggerty notes:

> [N]eo-liberalism arose to promote self-governance. It constituted individuals as integral to the management of the risks they face … [It also views] the individual as the rational manager of his or her own risk portfolio; he or she is expected to adopt a calculative attitude by becoming knowledgeable about risk. (2003, 193)

But, again, being 'calculative' and 'knowledgeable' is only permitted in relation to state-defined risks, and is illegitimate when it applies to the state *as a source of* risk. Such guides offer 'excuses that work' (Carroll 2002) or to 'Get ANY UK Speeding Ticket Cancelled Instantly!' (Speedcamerafine.com 2007), sources for protective equipment, legal 'get-outs' and tips for recognising different types of cameras with different capabilities: in short, all the necessary information for rendering oneself less vulnerable to successful prosecution for exceeding a speed limit. Such publications, including *UK Driving Secrets* (UK Driving Secrets 2010), *The Driver's Survival Handbook* (Thwaite 2008) and *Speeding Excuses that Work* (Carroll 2002), reinforce the idea of a collective oppressed minority, fighting back against an unjustified threat and supporting the image of the driver as the victim and as at risk. They are also generally written by an 'expert' such as a former police officer or 'multiply detected but never convicted' speeding driver.

The text of the advertisement used in the following example illustrates the feared consequences that can be drawn upon to motivate potential customers, phrased in the language of protection from an unfair threat:

> 'Shot by a speed camera?' What would YOU do right now if your license [*sic*] was revoked? How would you SURVIVE? It's more than just inconvenient. The impact is enormous. You'd lose your job, or revert to the dreads of public transport. You'd lose your social life. You'd lose income and social standing. You'd lose almost everything. And it could all happen just by being caught going at 90 MPH on an empty motorway! Are you angry? You should be! The whole system is a shambles, and you need to protect yourself. (UK Driving Secrets 2010)

The language of protection, threat, victimisation and even survival demonstrates just how effectively the proposed causal logic of the state has been inverted and is now deployed against itself.

Strategy 5: A 'black market' in points   Given that the majority of speed cameras currently in use take a picture of the rear of an offending vehicle, an opportunity has been created for responsibilised victims of speed enforcement to further minimise the impact of their victimisation. It is considered common knowledge within the speed enforcement debate that some drivers have nominated another individual as the driver of their vehicle when they have been caught speeding (see *BBC Inside Out* 2004). In doing so they have ensured that the other driver receives the fine and penalty points on *their* licence. Potential recipients of such points have been elderly or deceased relatives (with 'clean' licences no longer in use), spouses or partners who are less reliant on their licences (for example, the spouses or partners of professional drivers), foreign acquaintances (on the assumption that the authorities will not pursue such cases) or fiscally challenged students (for example, within clean licences but no car and no chance of getting one in the three-year 'life' of the points). There has even been a thriving internet-based market in selling points to other drivers (Millman 2006). Corbett et al. note that in their research:

> [t]here was a widespread 'folklore' about how other drivers avoided speed convictions. Many people thought the practice of passing penalty points to others was a common practice, and were able to identify groups who could be asked or paid to take points. Although reference was usually to 'other drivers', one respondent admitted to having passed points to other people on more than one occasion. (2008, 3)

Fleiter, Lennon and Watson also found references to the practice to be commonplace, both between friends and family and more 'systematically' where points were exchanged with strangers for cash (2007, 7–8).

Such a strategy had been adopted by several of the participants in this research. The following driver freely admitted to having transferred his points, much to the delight of the other members of his focus group:

> *Chris*: Well, I should have been banned a long time before I was, but I drive for my job and I got the wife to accept a few for me. She had six off me altogether but refused to have the last time, so I'm banned.

> *Sue*: Really? You did that? That's amazing!

> *Terry*: How on earth did you get her to do it?

> *Chris*: Well, she's not happy now because after my ban I'll have a clean licence again, but she's got my six for another two and a half years!

> *Terry*: My wife would say 'tough shit' if it was me. There's not a chance she'd have them off of me. Still, all kudos to you for doing it. (Convicted driver focus group)

It is impossible to calculate the likely take-up of this responsibilisation strategy on a national scale, but awareness of the potential for 'risky' drivers to evade disqualification in this manner apparently encouraged the development of cameras that provide an identifiable picture of the offending driver (BBC News Online 2006c). As a result, the enforcing technology is developed further to ward off attempts to render it ineffectual by responsibilised implicated risk instigators.

Strategy 6: Speed camera 'vigilantes'   A further extreme, and heavily publicised, response to the perceived risk posed by speed enforcement is the attacking of cameras by some drivers. The disabling of cameras by the use of firebombs, chainsaws and spray paint represents another method by which drivers can neutralise the threat as they experience it.

Some such instances of violent behaviour have been motivated by a detected individual's desire to destroy the evidence of their offending, and in doing so to protect the individual from the negative consequences that their detection would bring about, consequences that appear to centre around the symbolic security of the driving licence:

> A man who blew up a speed camera because he feared he would lose his licence after he was snapped speeding has been jailed for four months ... He claimed he was afraid he would lose his job if he was caught speeding. He already had ten points on his licence and thought he would receive a driving ban if more were added. (BBC News Online 2006d)

> A teenager who set fire to a speed camera that he thought had caught his car was banned from driving for a year. D, 18, a van driver ... feared that a speeding conviction would result in his losing his licence and his job. (*Daily Telegraph* 2005a, 1)

> A demolition worker who chopped down a speed camera with an angle grinder to avoid a £60 fine was ordered to pay £4,000 compensation yesterday and carry out 120 hours of community service ... Arrested at the scene he immediately told police he did it to avoid a speeding conviction which would have cost him his job. He was sacked when his employers found out what he had done. (*Daily Telegraph* 2003a, 3)

Responsibility for other acts of vandalism has, however, been claimed by a more organised group of drivers who go by the name of Motorists Against Detection (MAD). Their descriptions of themselves in their press releases, as well as the identity promoted in a telephone interview with their leader 'Captain Gatso', provide further evidence of the respectable self-identity discussed earlier, an identity threatened by speed limit enforcement by speed camera. The role of 'tax payer' is also promoted: 'Our members are responsible people in normal jobs, who, instead of paying into the system in the form of taxes are going to be faced with having to take out of it in the form of Income Support when they are stopped from working' (telephone interview with Captain Gatso).

The violent acts of vandalism perpetrated by MAD were presented as being the actions of a desperate but otherwise entirely well-behaved group of individuals, a fact that was perhaps designed to give them increased effect. The group presented themselves as responsible, honest and essentially harmless individuals who had been subjected to unwarranted and undeserved restriction. Their press release itself assured readers that: 'We are not criminals, just drivers going about our daily business and we are essentially law abiding citizens' (Captain Gatso 2002). In addition, Captain Gatso describes himself as 'late 30s, a family man with a respectable job' and his members as 'not boy-racers, they're all 35 plus, responsible people in normal jobs' (interview with Captain Gatso). Likewise, the organisation is keen to promote its normality in its press appearances: 'Our operatives are responsible people. Many are professionals with families who lead normal lives. These cameras are there to make money' (Captain Gatso, quoted in *Daily Mail* 2003, 43). The respectability of the 'normally law-abiding people' (MAD 2006) who made up the core of the group was also situated within a reference to their above-average skill levels as drivers. The possession of advanced driving qualifications is used to suggest a responsible attitude to driving, to counteract any suggestion of recklessness or dangerousness and to demonstrate the ordinariness of the group's members as opposed to their extremism or marginality:

> Capt. Gatso is a family man from north London. He's in his 40s, a professional he owns a BMW M3 and is a keen motorcyclist. He said: 'Most of the organising groups are just ordinary blokes with families who are sick of us heading towards a police state.' He added that the group's members were all good drivers, most have a professional driving qualification. (MAD 2006)

The self-promotion of some of these individuals as vigilantes or 'modern-day Robin Hoods' (interview with Captain Gatso) demonstrates that they understand themselves as both righting a wrong perpetrated against the public and acting to prevent the authorities from exploiting the public. In both cases, the state is reconstructed as the aggressor, and the illegitimacy of its intrusion is the motivating factor behind the action. Naturally, the government, understanding its use of cameras as a form of protection rather than a form of risk, then uses the criminal law to punish and incapacitate those individuals whose behaviour it construes as harmful. The underlying causal chains linking excessive speed to an increased risk of harm and their acceptance or rejection again dictate both responses.

Additionally, however, the speed camera itself has been redesigned to render it less vulnerable to attack in the perpetual cycle of risk and manufactured risk, having been described as 'the most assaulted pieces of street furniture in the land' (*The Guardian* 2006, 2). The 'ordinary camera with its radar mounted on a two-metre pole … is no longer muscular enough in an age of irate motorists and shadowy, direct-action campaigners such as Captain Gatso' (ibid., 2). It has been redesigned with a taller mounting and its wet film replaced with digital technology that cannot be destroyed; it also has its own CCTV systems to protect it from vandals. Such efforts can be seen as attempts to render the technology less vulnerable to its very own newly emergent risks, inflicted upon it by individuals who view it as a risk in itself.

Conclusion

Speed limit enforcement is seen to pose a threat to elements of the individual's personal and social identity, elements that have been rendered more vulnerable but simultaneously more important by the risk context that motivates the enforcement and in which that enforcement then takes place.

The consequences for one's moral, as well as social and economic, identity, seen to result from the enforcement of the criminal law against a mass activity, have been shown to result in a reassertion of the individual's essential law-abidingness and respectability. This functions as an attempt to deflect responsibility and blame for the harm of road death and injury. This deresponsibilisation takes place to the extent that individuals are able to reconceptualise the most pertinent risk as coming not from their own actions in respect of the speed limit but from the authorities' attempts to enforce those limits. The existence of the debate can therefore partly be explained in terms of the public protestations that make up drivers' attempts to construct their own case for being injured parties and not instigators and their campaigns to secure protection from, rather than be blamed for, risk.

In response to this perceived victimisation, individuals are seen to have resisted *responsibility* by instead becoming both *responsibilised* and *deresponsibilised* into taking action to reduce the consequences of this experienced risk. As such, some drivers can be seen to have become definers of risk problems in their own right and to have equipped themselves with the rationales and defensive routines that allow them effectively to reject the role of 'instigator' which the authorities have tried to assign them and to adopt more favourable identities.

Such responses are seen to result from the increasingly unpleasant consequences of accepting the role of 'instigator'. The resistance of this label is therefore more *attractive* and, given that a debate ensues about the accuracy with which the label has been applied, more *feasible*.

Chapter 6

Experiencing Automated Enforcement[1]

This chapter considers how the use of the speed camera itself as a technology of control contributes to the generation of objections to speed limit enforcement policies. It explores the way in which interactions with the camera technology specifically are crucial in forming a view of the experience, which is, in turn, key to an understanding of the concerns in evidence in relation to speed limit enforcement. Objections to the use of speed cameras are seen to centre on suggestions that enforcement is unfair and unjust. These claims will be considered here through the adoption of a framework of fair and just enforcement drawn from procedural justice literature.

This chapter begins by considering the increasing use of the 'techno-fix' (Haggerty 2004a, 494) in roads policing, before demonstrating the way in which such a mechanised system of enforcement, motivated by risk, is constructed in the case of speed cameras. The chapter then considers the experience of 'being techno-fixed' by the speed camera from a procedural justice perspective.

Techno-Fixing Road Risk

The shift towards risk as a focus and rationale for enforcement has produced changes in the way criminal justice systems operate (Sparks 2000; Hudson 2003). This has had consequences not only for the *type* of people criminalised (as Chapter 5 showed) but also for the *methods* that are used to bring about this criminalisation. One such change has been a detected increase in the use of 'techno-fixes', with such technologies offering the seemingly irresistible potential for cheaply and reliably identifying instances of so-defined risky behaviour (Seddon 2004, 16).

This shift in thinking towards a concern with the presence or absence of risk, and the change in detection technologies it enables, can be detected in a traffic context. Gaventa, writing for PACTS, describes 'a shift from the notion of enforcement as a concentration on deviance and individual offenders to an approach based on the policing, management and control of identifiable risks'. This leads him to adopt the hypothesis that '[n]ew enforcement technologies have assisted and accelerated movements in roads policing towards the policing and management of risk' (PACTS 2005, iii). Such new technologies included the 'impairmentometer'

---

1   Material from this chapter has also been published as Wells, H. 2008. The techno-fix versus the fair cop: Procedural (in)justice and automated speed limit enforcement. *British Journal of Criminology*, 48(6), 798–817.

able to measure 'reaction times and hand–eye co-ordination', as well as various types of impairment through alcohol, drugs or a combination of both (ibid., 64).

This preference for technological solutions has also been detected in wider circles, where drivers have been advised to 'brace themselves [for] [a]n array of technology ... that will radically change the way in which road safety laws are enforced':

> From a black box in the boot to 'alcolocks' on the dashboard, the car of the future
> is likely to be crammed with equipment designed to keep drunken drivers away
> from the wheel and slow down speeders. Motorists should brace themselves for
> more roadside cameras that will catch drivers who tailgate or do not put on their
> seat belt. (*Daily Telegraph* 2005b)

The two most recent pieces of road safety legislation to date also contain evidence of an official penchant for road safety technologies, featuring a large number of technological interventions designed to impact on a variety of problems and with a view to achieving casualty and crime reduction targets (DfT 2005f, 2011). These include, in the 2005 document, 'alcohol interlocks' which require drivers to provide a sample of breath that contains a legal blood-alcohol limit before the car will start, Automatic Number Plate Recognition (ANPR) systems which detect the presence or absence of vehicles on certain key databases, such as insurance, road tax and MOT databases, and the further use of speed cameras. A further four proposals have also been made which would increase the effectiveness of existing techno-fixes.[2] The 2011 Strategic Framework for Road Safety also continued official interest in road safety technologies, with the intention to type-approve portable evidential digital breathalysers (allowing evidential quality readings to be taken at the roadside), to authorise the use of a drug-screening kit for use on suspected drug drivers and to consider 'a new drug driving offence if the current offence of driving while impaired can not be made to work more effectively and the research on impairment and technology on detection allows this' (DfT 2011, 10).

The result of this enthusiasm for technological solutions to roads policing problems is that the drivers are now more likely than ever to encounter authority in the form of a machine, often a camera. Because such interventions are incapable of measuring any factor other than the 'risk factor' that has been used to justify their use in the first place, such technologies have been identified as offering the welcome potential for guaranteeing entirely non-discriminatory enforcement: 'It is the first time in human history that we have the opportunity to experience forms of control that do not take into account any category of social division. Age, sex, race, beauty and attire are irrelevant and, what is equally important, guaranteed to

---

2   Banning speed camera jammers and detectors, the improved evidential value of roadside breath tests, improved enforcement of tachograph restrictions and measures to combat 'clocking' of vehicles (DfT 2005f).

be so' (Lianos and Douglas 2000, 108). The techno-fix is therefore guaranteed to be 'fair' in one sense, and the potential introduction of technological control into the roads policing context has been heralded as offering the potential to 'render obsolete the litigation, public criticism, and academic critique' associated with human officers carrying out roads policing duties (Joh 2007, 216).

However, the experience of interaction with speed cameras as a form of techno-fix is commonly described in terms suggesting that such guarantees have not produced a system that is experienced as the positive non-discriminatory encounter anticipated. Instead, such encounters are often described as 'unfair' or 'unjust'. Having considered the techno-fix in more detail, these responses are then explored below using a framework of analysis drawn from the concept of 'procedural justice'.

The Techno-Fix and its Assumptions About Risk

Techno-fixes work on assumptions about risky/non-risky, safe/unsafe behaviour and detect actions that fall the wrong side of the line such that an intervention, often in the form of a sanction, can be applied. This simplified differentiation between acceptable and unacceptable behaviour (the identification of the 'good' from the 'bad', we might say, given the discussions in Chapter 5) can be easily identified in many road safety interventions. A proposed camera that could catch 'tailgaters' (*Daily Telegraph* 2005b), for example, must operate on a generalised, decontextualised, interpretation of 'safe' and 'dangerous' distances between cars, while in the case of drink driving, moral and immoral behaviour is represented by a blood-alcohol limit of a certain level. The 'impairmentometer' which has been suggested at various points and been picked up by the media (Times Online 2004; PACTS 2005) would also, presumably, create such distinctions for eyesight, tiredness, hand/eye coordination or whatever other dangerous condition it promises to measure. Speed limits are the obvious point of differentiation on which speed camera systems are based, with speeds in excess of the speed limit in place at a particular location interpreted as increasing the potential for a crash to occur at that location. This means that a change in risk levels is taken to occur at that particular level, at the round figure of 30, 40, 50, 60 or 70 mph, regardless of the weather conditions, traffic levels or time of day to which that limit is applied.

Following the moment of detection, and given that an act's 'riskiness' is the justification for deterring the behaviour, the legal framework subsequently adopted to underpin enforcement need only be constructed in a way that determines that this faulty behaviour occurred, not that it was *deliberately* engaged in. Both deliberate and accidental behaviour are therefore legitimate targets for enforcement action justified on risk grounds, meaning that strict liability legal principles can be used. Complicating and unnecessary issues such as intent, culpability and mitigation are removed both through the use of an acceptable levels framework enforced by the techno-fix and through the use of strict liability principles. Individuals and their behaviour are therefore reconceptualised as either triggering or failing to trigger the techno-fix. Costly and time-consuming court appearances are unnecessary and are designed out of the process. We can see this process in operation in both the speed camera context and in a variety of other offences that are now thought to be suitable for the 'fixed penalty'.[3]

There are clear economic and efficiency gains to be made by viewing problems in risk terms. The speed camera can then operate as a stand-alone detection technology, producing sufficient evidence to instigate prosecution procedures without the need for human corroboration and thereby reducing both costs and labour requirements. Unlike the human it has replaced, the automated nature of speed camera technology means that it can be on duty constantly, does not tire towards the end of a shift and operates to the same standard of accuracy in its measurements every time it is activated. This produces a large number of identically perceived infractions for which identical punishments must be dispensed. The fixed penalty system that underpins enforcement is therefore both viable, given the standardised response that speed limit infringements are deemed to warrant, and necessary, given the vast numbers of cases that are produced through the use of automated detection technologies. The techno-fixing process is thus constructed in a way that allows for the efficient and economic detection of behaviour defined as risky and the efficient and economic dispensing of punishments for those who have engaged in such behaviour. Despite the increasing use of techno-fixes, strict liability legal frameworks and fixed penalty disposal mechanisms, relatively little attention has been paid to the changes that this has brought about for the individual whose behaviour falls the wrong side of the line and who therefore becomes subject to this reconfigured justice process. Given that exceeding the speed limit is one offence that is characterised by all these reconfigurations (automated detection, acceptable levels, strict liability, fixed penalty) and is, furthermore, something very familiar to a large and increasing section of the driving population, speed cameras make a particularly good case study through which to explore the social experience of enforcement.

Injustice, Unfairness and the Speed Camera

Despite the use of supposedly 'fair' and neutral enforcement methods, the experience of interaction with speed cameras as a form of techno-fix is commonly described as 'unfair' or 'unjust':

---

3   The appeal of fixed penalties means that they have increasingly been promoted as the appropriate intervention for a variety of unacceptable behaviours, including unauthorised parking, littering, smoking in enclosed public places and over-filling dustbins, as well as criminal activities such as graffiti, criminal damage and being drunk and disorderly (Home Office 2000). Fixed penalties have also been promoted as a major tool in the armoury of Community Support Officers (Home Office 2006).

I just feel cheated out of my money, I really do. I think the whole thing is just so unfair it makes me very angry. I am not an angry person, but the injustice of the whole thing really gets me going. (Female driver, early 40s, experienced driver focus group)

Ray: They put a speed camera in on the A34 near Tesco and they put it right in behind the 30 mph speed sign …

Dave: … and they had to take that one out, yeah, you're right.

Ray: … they had to take that one out. People said, 'That's not fair! You're not being fair!' and I can certainly see their point. (Male drivers, late 50s, professional driver focus group)

What they've done to me, banning me, I mean they've just carried on with a bit of 'justice' from 20 or 30 years ago. It seems to me to be totally unfair. Twelve points always used to be the limit when it was almost impossible to get stopped that many times by a policeman, but now you can do it in one journey and that's it – banned! (Banned male driver, mid 50s, convicted driver focus group)

Additionally, general feelings of underhandedness, sneakiness and unfairness can be detected in the coining of the terms 'speed *trap*' and '*scam*era'. A poll conducted in 2003 also found that over 80 per cent of motorists questioned felt the government treated motorists 'unfairly' (YouGov 2003).[4]

Cleary, then, 'fairness' is dependent on more than simply the removal of the potential for discrimination based on age, gender, race, class or any other factor. One body of research that suggests a broader definition of 'fairness', and as such offers the potential to advance discussion in this area, is that which explores the notion of 'procedural justice'. This chapter therefore considers additional criteria that make up 'fair' policies and explores their presence or absence within speed camera-based enforcement systems. As such, a framework for 'fair' enforcement is proposed and the relevance of the use of a 'procedural justice' framework is demonstrated, before being considered in relation to the NSCP and subsequent use of speed cameras.

Procedural Justice

The allegations of 'unfairness' and 'injustice' levelled at speed cameras, above, should not, it is suggested, be viewed as the inevitable complaints of a group of individuals who are being punished and prevented from behaving as they wish.

---

4   In response to the question about how the government treats motorists, 84 per cent of respondents selected the option 'unfairly' from a list that also included 'too generously', 'about right' and 'don't know'.

According to the procedural justice perspective, if punishment is carried out in particular ways, it need not be a delegitimising experience for those who receive it. Punishing policies even have a chance of being seen as *positive* experiences when they are designed in such a way as to be regarded as legitimately and fairly *motivated*, and subsequently *operationalised* in ways that are also experienced as fair, just and legitimate.

In the first instance, fair and just policies are seen to be those brought about by 'trustworthy' authorities (MacCoun 2005, 182). This perception is based on 'a belief that the authority involved is not behaving in a biased or self-interested manner', and that it is thus intervening in 'good faith' because of a genuine belief about the likely positive effects of its actions (Tyler 1990, 117).[5] Judgements as to the trustworthiness of the authorities are, however, based on the *operationalisation* methods chosen by the authorities, given that motivation cannot be directly observed. Such implementation methods must be 'consistent' (Tyler and Lind 1988, 131) and demonstrate 'neutrality, lack of bias, honesty, [and] efforts to be fair' (Tyler 1990, 7) as well as 'impartiality' (ibid., 117).

MacCoun has also noted the importance of the presence of 'signals that convey our social standing' in procedurally just policies (2005, 182), while Tyler, too, notes that 'respect for citizens' rights', 'politeness' (1990, 7) and the 'consideration of one's views' are crucial antecedents of a procedurally just experience (ibid., 117). Further elements of this experience include the ability to 'voice' (Tyler and Lind 1988, 170–72) and achieve a degree of 'process control' throughout the experience by being given the opportunity to vocalise a point of view and have input into the decision-making process (Tyler and Huo 2002, 88). Such a framework for analysis is considered to be particularly important to this enforcement context for a number of reasons.

Policies that demonstrate these qualities allow the authorities to carry out their duties of punishment, reprimand and control without damaging the respect with which they are viewed: 'According to a normative perspective, people will be concerned with whether they receive fair outcomes, arrived at through a fair procedure, rather than with the favorability of the outcomes ... people want justice from police officers and judges, and evaluate those authorities according to whether they get it' (Tyler 1990, 5). The ability to inflict punishment without damaging legitimacy is dependent on the procedures that are used to dispense such 'unfavourable outcomes': 'If unfavorable outcomes are delivered through procedures viewed as fair, the unfavorable outcomes do not harm the legitimacy of legal authorities' (ibid., 107). Such legitimacy is, it can therefore be suggested,

---

5   Tyler and Lind also note that 'policies should be presented to citizens as having been developed in a fair manner' (1988, 163), something which, in the context of regulatory offences, must relate to the existence and communication of a persuasive rationale and research base supporting the intervention. This relates to the fairness with which scientific methods have been used to reach policy decisions, with it being necessary that the public appreciates that policies are the fair and just result of the policy-making process (ibid., 154).

particularly important in respect of regulatory offences such as speeding, given the likely absence of normative commitment to such rules, and the consequent reliance on the connection between the action and the harm in justifying such unfavourable outcomes. Individuals are likely therefore to choose to accept or reject unfavourable outcomes in respect of regulatory offences following an assessment of the procedural justness with which they were delivered.

The methods chosen to detect speed limit infringements are, furthermore, considered to be particularly significant given that they target a population with little practical experience of the criminal justice system, with 'police officers' or 'judges' (ibid., 5): '[M]ost people have little contact with *any* legal authority. Therefore each personal experience is likely to be memorable and to play an important role in shaping a person's views about the law and legal authority' (Tyler and Huo 2002, 131). Corbett notes that roads policing is now 'the most common means of police-initiated contact with citizens' (2008, 132) and suggests that it '*is* the public face of the police for many citizens' (ibid., 131). The qualitative nature of these interactions is therefore of particular importance from a procedural justice perspective. Given that the likelihood of punishments being accepted is greatly increased by dispensing them in procedurally fair ways, a large number of people are potentially going to be alienated by policies that pay insufficient attention to these concerns.

Furthermore, procedural justice authors have noted that 'socially connected' individuals 'base their judgments more strongly on the degree to which they receive procedural justice' in encounters with authority (Tyler and Huo 2002, 124). Given that this policy implicates individuals who consider themselves respectable and socially responsible in unfavourable-outcome encounters with authority, the importance of procedural justice issues is clear in relation to this policy perhaps more than any other that may focus on more traditional offending groups.[6]

The procedural justness of the experience is, additionally, found to be of particular significance when the authorities intervene unsolicited to impose restrictions (ibid., 56). Given that speeding is a seldom-reported offence, policed proactively and without the need for public complaints, it is a particularly important candidate for procedurally just enforcement processes.

Finally, but of crucial importance to this work, this theoretical framework also identifies the potential consequences of operating procedurally unjust or unfair policies. These take the form of non-compliance with the law and reduced cooperation with authority, produced via a decline in levels of their perceived legitimacy (Sunshine and Tyler 2003, 514). These outcomes are not, furthermore, confined to support for the policy in question, but extend beyond this to pollute the relationships between the state and society at other points of intersection (Tyler and Lind 1988, 149). In Partnership arrangements, such as those through which

---

6   Less socially connected individuals are, it is proposed, more concerned with distributive justice (outcomes) than with procedural justice (fairness) (Clayton and Opotow 2003, 303).

speed limit enforcement has been pursued, it cannot be discounted that such effects could therefore also damage public relationships with other agencies involved in speed limit enforcement.

Procedural justice issues are therefore of considerable potential importance in relation to speed limit enforcement.[7] However, an assumption has always been made that procedurally just systems will be generated within the context of inter*person*al encounters (see, for example, Tyler 1990, 134; Sunshine and Tyler 2003, 533). The antecedents of consistency, neutrality and impartiality are considered in the context of the respect, politeness and opportunities to voice provided by systems that inevitably centre around scenarios in which humans interact. In the case of the techno-fix, such inter*person*al interaction does not occur. Authority is initially represented by a machine rather than a human being, and subsequently distanced from the problematic individual by postal fixed penalty systems which replace court appearances. Such a qualitative shift in the nature of unfavourable outcome encounters is thus a further justification for considering the compatibility of procedural justice with the techno-fix.

The following analysis will therefore suggest some reasons for negative responses to speed limit enforcement which relate specifically to the types of enforcement methodologies and methods of implementation made possible by a dominating concern with the minimisation of risk. Their in/ability to offer procedural justice to those who fall within their regulatory gaze will frame this discussion. As such, additional antecedents over and above, and in some cases in place of, those elements of fairness guaranteed by techno-fixes will be considered. An explanation for the absence of 'fairness' in guaranteed non-discriminatory encounters will be proposed in relation to the stages of the techno-fix outlined at the start of the chapter. First, the use of the 'acceptable level' is considered, before the use of strict liability and fixed penalty processes is explored. The specific role of the camera as a techno-fix is then considered, and finally the impact of these methods of operationalisation on perceptions of the motives of the enforcers is discussed.

'Acceptable Levels' and Consistent but Unfair Enforcement

A common objection to speed limit enforcement relates to the centring of the policy around the notion of the speed limit as the point of differentiation between problematic and unproblematic behaviour. This method of differentiation offers the apparent potential for entirely consistent, neutral and impartial enforcement, applying the same controls to all drivers in all circumstances, but is nonetheless experienced by some as unfair. Objections centre around the idea that limits somehow always represent an accurate judgement of acceptable speed, regardless of other variables which change constantly and affect the 'actual' (if such a status

---

7   See Soole, Lennon and Watson (2008) for further engagement with the related issues of legitimacy in the speed camera enforcement context.

could be known) level of risk posed in a specific context. These contextual issues include, for example, the driver, the time of day, light levels, weather conditions or the presence or absence of pedestrians, and are not, indeed *cannot*, be reflected in fixed limits. The very factors that make the use of limits attractive from 'risk' and 'techno-fix' perspectives do therefore offer the potential for damage to the perception of the policy as procedurally fair or just. The system's ability to utterly decontextualise an offence, recording only that an offence took place and that a penalty should be assigned for a breach of a proxy-measurement of risk, can be considered to disregard and disrespect differences that some drivers believe *should* matter.

Limits and disrespect of offence context   Although a vital aspect of this kind of automated road safety intervention, limits are seen by some drivers to be inappropriate in many circumstances. This is seen to lead to both the unnecessary and unfair restriction of drivers in some contexts and their insufficient protection in others. The overly consistent application of limits to situations and contexts that are anything but consistent is experienced, by some, as unfair:

> *Pete:* Eighty [mph] in some situations is safer than 30 in others …

> *Dave*: Yeah, that's true enough.

> *Pete*: … and there's a well-used saying that 'there's nothing wrong with speed so long as it's used in the right place and at the right time', and that's quite true. (Professional driver focus group)

> I know there's good reason for certain speed limits in certain areas, but I mean there is also a good argument to suggest that speed limits, you know 24/7, is nonsense. Because OK, outside a school 20 mph, got no problem with that, but when it's school holidays, why keep it at that level? Why not have sort of variability to do with time of day and congestion? (Male, mid 40s, convicted driver focus group)

This view is supported by research by Blincoe et al. who offer a similar quotation from a driver who, having staked his own claim for being regarded as an expert by noting that he is 'an advanced driver', asks, 'With nobody on the road or pavement and good weather conditions, why slow down!?' (Blincoe et al. 2006, 376).

From a procedural justice perspective, it is notable that not all limits are seen as unfair or inappropriate. Their 'justness' is dependent on the *context* to which they are applied and the accuracy with which they seem to apply to it. Limits that aim to reduce speeds in areas outside schools, for example, were commonly supported as 'fair', often in connection with an acknowledgement that these were locations where increases in speeds could readily be imagined to increase levels of risk to a very obvious potential and deserving victim.[8] In such situations, the stated road safety motive of the authorities can be trusted because the motive appears consistent with the method of operationalisation chosen, and the speed camera is accepted as a road safety intervention (and see Soole, Lennon and Watson 2008).

Limits and disrespect of offender context   In addition to the consistent application of limits to inconsistent *situations*, speed enforcement was viewed as unfair by some, given that it consistently applies the same limits to all *drivers*. Duff notes that, in the context of strict liability laws, there will always be some people for whom regulation based upon averages is unnecessary, because a proportion of the population will be above average, giving speed limits as an example of this phenomenon:

> [T]he driver who knows that she can drive quite safely and competently at an illegal speed or when over the limit for alcohol, do[es] not expose anyone to an unjustified risk of harm; and [her] conduct is wrongful, if it is wrongful, only because the law defines it as criminal – which is to say that the offence ... is *malum prohibitum,* not *malum in se.* (Duff 2002, 102–3)

Similarly, Beck has noted that in relation to toxins, pollution or radiation acceptable levels are established on the basis of the impact of certain actions on the *average* individual. Such impact then comes to dictate the level to which everyone is exposed (1992, 64–9) and can be seen as unnecessary for those who believe they have an above-average profile.

Strict liability laws based on acceptable-level principles therefore inevitably restrict some people who perhaps might not be as in need of regulation (or at least consider themselves to be so) as others. Prior to enforcement methodologies of an automated nature, enforcement of traffic law was seen to 'discriminate against those law violators who experienced accidents' (Ross 1973, 77). Only those who were involved in crashes were likely to be punished, filling the courts with crash-involved drivers and thus reinforcing the idea that bad drivers were drivers who drove illegally *and caused crashes*. With current enforcement methodologies, however, the opposite case is true, with those drivers detected and prosecuted for speeding being those who have demonstrably not been involved in an incident of any kind on that occasion. Detection by a speed camera instead testifies to the absence of any 'actual' harm being caused, given that if a crash had been caused at that moment, alternative legal disposals such as charges of 'careless' or 'dangerous' driving would have been used. Enforcement in a pre-risk climate could be seen to more accurately detect dangerous drivers because their involvement in accidents was proof of their dangerousness.

---

8   'Deserving' in the sense of victims more deserving of sympathy for their plight, not deserving *of* their plight.

Although many drivers chose to identify themselves in terms that reflected their belief in the importance of their prior driving record, in the new risk-motivated enforcement climate the fact that a driver has 'been driving for 40 years with a completely clean licence' (female, late 60s, Speed Awareness Course) is able to claim that 'I got my licence in 1960 and this is the first time I've got done for anything' (male, 60s, Speed Awareness Course), or was even 'Lorry Driver of the Year three times' (professional driver, Speed Awareness Course), is irrelevant if their behaviour failed to meet the required standard at the most recently encountered checkpoint represented by the speed camera. There is no 'no claims bonus' when it comes to speed cameras.

Given that as many as 90 per cent of drivers do indeed consider themselves to be better than average (Delhomme 1991), 90 per cent of drivers are likely to consider themselves unnecessarily inconvenienced by laws designed to restrict the behaviour of *average drivers*, based on the levels of risk that *they* would cause in any specific set of circumstances (and see Musselwhite et al. 2010, 5). Given this statistic, traffic regulations that are based around strict liability premises and which use the logic of the acceptable level are therefore perhaps particularly liable to accusations that they fail to give due consideration to individual drivers' abilities. This 'self-enhancement bias' has been noted by other authors (Groeger and Brown 1989; McKenna 1993; Walton and McKeown 2001; Musselwhite et al. 2010), with Walton and Bathurst arguing that this 'seemingly universal phenomenon' (1998, 1) 'leads to drivers believing that traffic rules, such as speed limits, only apply to less skilled drivers, rather than themselves' (Walton and Bathurst 1998, cited in Blincoe et al. 2006, 372). Corbett's research also found that being caught speeding was experienced as 'an unnecessary rebuke or unwanted harassment, as they are not the cause of the speed-accident problem' (2000, 326), while her earlier research (with Simon) suggests that this sense of superiority in driving skill helps to protect drivers from challenges to their identity as explored previously:

> [T]hrough feeling in control, many drivers were able to perform an unlawful act without regarding it as criminal or immoral ... It appeared that provided they felt in control during the manoeuvre they were apt to dismiss the notion of a traffic offence being immoral, since harm was thought unlikely to result and would in any event be unintended. Thus this attitude allowed them a clear conscience. (Corbett and Simon 1992, 545)

By viewing the use of speed limits from this perspective, we are also able to explain the apparently anomalous finding that, while about 74 per cent of drivers agreed that speed cameras should be used to enforce speed limits (Transport2000 2003), 85–99 per cent of drivers then admitted to breaking those speed limits (Corbett 2003, 111). If drivers consider that the majority of other drivers are less skilful than themselves, they are likely to approve of a measure that limits the amount of risk those other drivers can pose *to them*. They are, however, unlikely to consider that their own behaviour is harmful or worthy of restriction and therefore choose to break speed limits that they desire other drivers to adhere to.[9] Research has shown that offending drivers do typically accept that there is a link between accidents and other drivers' high speeds, but not their own (Corbett and Simon 1992, 546; Corbett et al. 2008, 35).

The enduring sense of superiority of skills evidenced by some drivers would also, it might be suggested, have consequences for the public acceptance of Parker and Stradling's formula 'Violation + Error = Crash' (Parker and Stradling 2001). The formula, used several times to underpin policy making in this area (see DETR 2000a; TLGR Committee 2002), suggested that speeding (as an example of a violation) is combined with drivers' mistakes to lead to collisions. The likelihood of a driver accepting that their violation will be combined with some error on their part and thus lead to a collision is reduced by any belief in their own above-average skill levels and thus their self-perceived invulnerability to 'error'. Essentially, the self-perceived above-average driver can therefore object to being punished or restricted as a 'false positive'. The denial of the significance of relevant differences may, once again, be understood as disrespect caused by the system's inability to recognise context.

Rather than producing a procedurally just experience, the antecedent element of 'consistency' can be interpreted as providing insufficient respect for individual circumstances and the specific context in which each offender offends. Consistency is therefore *un*fair when it leads to the disrespect of differences that are thought to be significant in determining whether or not risk was really caused.

As such, the procedural justice antecedent element of 'respect' is seemingly conveyed through discriminatory experiences: discrimination that is understood in this context to relate to the appropriately different treatment of qualitatively different circumstances and characteristics, and which is incompatible with systems based around the techno-fix. Inconsistent treatment, it seems, is experienced as fair when it allows for genuine differences in context to be respected.

Fixed Penalties, Strict Liability and Disrespect Through Denial of Voice

The contextual issues raised in the above analysis suggest that each offence, rather than being a decontextualised incident, has a story attached. One aspect of this story concerns the construction of explanations as to why enforcement and punishment are not necessary or valid in any particular individual case. Participants in this research frequently supplied their own reasons why their criminalisation was unjustified, why it was unfair and why acceptable levels were unacceptable in their specific case. The formulation and communication of such

---

9   See Stradling et al. (2003), for example, where general opposition to breaking speed limits is presented alongside data showing that the majority of people still do break them themselves. We cannot know from this type of example whether the respondents were thinking in terms of their own permissible behaviour or that which they considered permissible for other drivers.

excuses, reasons and mitigation for apparent incidents of offending behaviour is essential to the maintenance of a coherent identity in the face of challenges from authorities (Sykes and Matza 1957/2001, 209).

In the majority of criminal cases, the courtroom provides the venue for excuses, mitigation and rationalisations to be heard in front of a representative of the criminal justice system that has sought to criminalise the behaviour. Again, such opportunities for control and participation have been shown to be vital to procedurally just policies. Such participation was measured by Tyler and Huo in terms of participants' sense that they were able 'to say what was on [their] mind', 'able to make [their] views known' and were given 'an opportunity to tell their side of the story' (2002, 194). Furthermore, '[t]he perception that one has had an opportunity to express oneself and to have one's views considered by someone in power' is seen to play 'a critical role in fairness judgements' (Tyler and Lind 1988, 106). Once again, however, procedural justice advocates assume that such opportunities are made available within a courtroom scenario and through the interaction of a judge and a defendant:

> [T]heories of procedural justice suggest that people focus on court procedures, not on the outcome of their experiences. If the judge treats them fairly by listening to their arguments and considering them, by being neutral, and by stating good reasons for his or her decision, people will react positively to their experience, whether or not they receive a favorable outcome. (Tyler 1990, 5–6)

Given that fixed penalty and strict liability arrangements effectively remove this opportunity for all but a minority of defendants, this human interaction and the sense that the case has been, effectively, 'heard' is absent. As O'Malley notes, '[t] he judge has disappeared' (2009b, 77). Some drivers identified this absence of any opportunity to voice as a significant omission in the current system:

> There needs to be some way you can tell them that the signs were bad, or that there was no one else within ten miles of you. The camera can't tell that obviously, but surely there must be some way that you can tell them things like this? (Male, early 20s, new driver focus group)

> I wanted to tell someone! To say 'Look, this is ridiculous!' The camera was right inside the 30 limit, I mean virtually level with the signs and I just didn't have the chance to slow down. There must have been dozens of us done just the same. But who do you tell? Is it the council? (Female driver, 40s, experienced driver focus group)

Such protests are suggestions why the contexts of particular offences make them invalid, less serious or excusable and consequently why punishment is not deserved or fair. Opportunities to offer these explanations, which *are* offered at other stages in the system, are therefore particularly cherished:

Laura: You know when you used to get sent the form there was a section that asked for any reasons you were speeding, and I always had that, but the last one or two that I had didn't have that, there was nothing to say 'why', so they'd obviously ... well, you never thought for a moment that they'd take any notice but at least it was there so you could *say*.

Sue: You feel as though you are *doing something* if you write in. It said on my form 'no appeal'!

Laura: But what I mean is that they've obviously done away with that as well, which I guess proves they never read it.

Sue: I know what you mean. It was like at least they cared about stuff like that. (Convicted driver focus group)

To these drivers, the opportunity to communicate important aspects of the context of their offences to the authority that sought to criticise and punish them was valued in and of itself, regardless of its perceived potential to change the outcome of their case. The authority that offered opportunities to communicate such comments and concerns was considered to 'care' about context whereas the authority that denied such an outlet was viewed less positively. This suggests that the same outcome (punishment) could be rendered more acceptable if accused drivers were given the opportunity to 'voice'. Such a view is confirmed by other research (Tyler 1990; Johnson 2004; Shephard Engel 2005).

That some drivers miss this outlet for telling their story is also indicated by the way in which magistrates' courts, fixed penalty payment offices and Safety Camera Partnerships receive frequent calls from drivers who simply want to explain the circumstances of their offending to another human being.[10] The idea that the court, rather than the roadside, is considered to be the appropriate venue for guilt to be determined is clearly in evidence in some drivers' criticisms of the remote fixed penalty system:

Sue: On the form you should at least be able to send in a comment about the stretch you were done in, if the limit was correct or not.

Laura: You've had your chance to explain whether or not they take any notice or not.

Terry: There are penalties out there for people who drive really dangerously, all we have to do is hand them out appropriately. And that's what you have a judge and a jury for, for the particular sentence, which is based on all the

---

10   Based on personal experience of working in, and conducting observations in, these venues.

circumstances of that particular case, or accident, and you get the chance of an appeal as well. There are checks and balances in the *proper* system. (Convicted driver focus group)

Seemingly, accused drivers seek out human contact in order to 'voice' their explanations, but also potentially, given the analysis above, to reinject some (positively viewed) discrimination back into the system and to have considerations such as their above-average skilfulness acknowledged. The absence of such opportunities within the standard fixed penalty process is viewed as unfair given that it *formalises* the authorities' lack of interest in the context of the offence and the offender. As O'Malley notes, a kind of partial, dividualised, 'simulated justice' is the destiny for most and 'we are only hailed as individuals either when we choose to resist and demand to be treated as individuals or when we cross a risk threshold'. Such opportunities are too rare to be considered to represent what justice 'really is' in this context (2010, 805).

Significantly, the perceived denial of individual circumstances, excuses and contexts when the 'law-abiding majority' encounter authority is contrasted with a perceived 'culture of excuses' operating with regard to 'real crime':

> They never send burglars to jail any more, there's always some reason why they get let off, but break a speed limit and the whole weight of the thing comes down on you. It doesn't matter if *you* are poor, or got tortured in some African country, or come from a big family where Daddy didn't love you, does it? (Male driver, late 50s, professional driver focus group)

'Real criminals' are seen not only to be allowed the opportunity to have their circumstances respected, but even to be encouraged to use any imaginable excuse for their genuinely harmful offending. In contrast, therefore, the fixed penalty system and strict liability status of speeding denies drivers the opportunity to exercise what, to them, are entirely valid reasons why they should not be cast as offenders or branded as risky.

'Disrespect' of the respectable   The use of automated enforcement technologies has been shown to have altered the nature of the interaction between the criminalising authorities and those they wish to criminalise, and as such the speed camera's influence is discernible throughout the enforcement process. However, the speed camera is also understood as offering a specific kind of disrespect to some of those it watches over.

Speed cameras are capable of receiving only certain types of input and generating certain types of output. They are not, in short, capable of meaningful interaction with those over whom they exercise control and they are, as a result, the on-the-ground manifestations of the lack of interest in context that the former structures have brought about. MacCoun has noted the importance of 'signals that convey our social standing' in procedurally just policies (2005, 182), and the 'respect' with which individuals are treated has also been seen to be of significance (Tyler 1990, 7). The use of cameras to enforce speed limits is seen by some drivers to convey signals that are insufficiently respectful of their social status.

In the following example, a driver singles out the 'facelessness' and inaccessibility of camera-based enforcement as a factor in creating the resentment towards the policy as a whole:

> It's an emotive issue and I think that it's because it's all about control and it's about an authority, it's about a faceless sense of authority who we don't see but we receive a letter from, telling us to moderate our behaviour. I think that's why it creates such a bone of contention and why people see it is unfair. No one likes to be criticised, but more than that, no one likes to be embroiled in the criminal justice system, I think. The idea that in one moment, one over-excessive push on the accelerator and, despite being a respectable, law-abiding bloke, you enter into the criminal justice system and become stigmatised thereafter, I think creates a fear, and that fear creates a resentment and an anger towards this faceless authority, towards this *machine* that has judged you. I mean, I don't even know myself, transport partnership, camera partnership whatever. I mean, is it the Inland Revenue who gets the money, who sends the bill if you get a bill? I don't know. But it's a bad thing. (Male, 40s, experienced driver focus group)

As such, the inaccessibility of the enforcers and the faceless judgements that their deployed technologies make is viewed as contributing to the sense of injustice with which speed cameras have been met, particularly for some 'types' of driver.

For 'respectable, upstanding members of the community' (female driver, mid 60s, convicted driver focus group), encounters with the criminal justice system have historically been characterised by the 'niceties' of due process, justice and, it is suggested, more sympathetic and respectful treatment by human enforcers (Stenson 2000, 19). Since the widespread automation of speed limit enforcement, however, this differential treatment has been denied to those who previously felt they benefited from it: the respect*able* are experiencing disrespect given that 'status' cannot be effectively 'deployed' in interactions with non-human technologies (Lea 2002, 15). All the system is interested in now is potential risk, and its robotised enforcers judge all potential threats as equally likely to offend and thus deserving of the same level of suspicion and restriction: 'Everyone is presumed guilty until the risk profile proves otherwise' (Ericson and Haggerty 1997, 42).

The reassertion of one's essential respectability and law-abidingness is perhaps an understandable challenge to efforts to reclassify on the grounds of risk, with such reassertions made particularly necessary by a detection method that entirely devalues such identities and makes them irrelevant. Being a law-abiding, respectable or upstanding citizen (all important aspects of the identity of some drivers as the previous chapter showed) are all references to *previous* behaviour, to worked-at and worked-for status, and to established and maintained credentials of worthiness which are valued by their owners. They are, furthermore, representations of individual, rather than dividual identities (Deleuze 1995). But these complete identities are found to be of no use in the current context of speed enforcement in which each problematic event is 'untouched by human hand' (O'Malley 2009b, 77) and where worth is instead judged against limited (and apparently disputed) criteria that are measured only at proliferating checkpoints.

Unbiased enforcement can therefore be experienced as *too* fair, in that it does not discriminate enough for people who are used to discrimination working in their favour. 'Justice' and 'fairness' are, for such drivers, not simply about consistency in the application of legal rules, but also necessarily involve some sense of case specificity that allows for the circumstances of each case to be taken into account (Hudson 2001, 145). Hudson refers to this specificity in terms of the feminist observation that 'treating people the same is not necessarily treating them all equally in the sense of giving them equal consideration' (ibid., 164). However, it seems that it is the traditionally power*ful*, not the traditionally power*less*, who are seeking to exploit case specificity to their advantage in this context.

Ulterior motives for enforcement: Motive-based (mis)trust As a second element of the formation of procedurally just policies, the perceived motives of the enforcers and the rationale underpinning their actions are also of importance, as indicated at the outset of this chapter. Such policies are, in addition to being implemented in 'fair' ways, also seen to be those that have been created by 'trustworthy' authorities (MacCoun 2005, 182). The absence of 'bias' and 'self-interest' (Tyler 1990, 117) has been shown to be an essential aspect of the way in which the policy-making process is perceived, and it is these two aspects that are considered in this section.

This motivational aspect of procedurally just policies is clearly linked to the methods of operationalisation considered in depth above. Observations about the methods employed by an authority are, it is suggested, used to form such judgements about that authority in terms of its trustworthiness and its motivations. Given that these motivational qualities are not *observable* phenomena, the *actions* of enforcing authorities are used by individuals to make *inferences* about the motives of those who seek to regulate their behaviour (Tyler and Huo 2002, 52 and 61):

> *Alan*: I think there's certain occasions where they are in the wrong place, know what I mean? Sometimes they're out in the middle of the countryside, know what I mean? I can understand schools and built-up areas, stuff like that, towns and stuff, but I mean we had one driver nabbed … and he was out in the countryside. I think there might have been one farm in about a two-mile stretch, and there was one [camera] there. I just can't see … unless there's been accidents there, but I just can't fathom out why it was there.

> *Simon*: The more that get put up, the more people think that they're just to bring in revenue. (Male drivers, professional driver focus group)

I travel the roads over the moors and they claim that between four and five bike deaths happen per year and the numbers of speed cameras on those roads altogether is three, yet on the A34 into Stafford there are 32. And the road over the moors is a demanding road with twists and turns and farm entrances and things like that, and it's not an easy road to drive. So even though I'm aware that there's a road safety element to some extent, I'd put that at 10 per cent with 90 per cent as revenue generation. Very, very sceptical. That doesn't mean to say I'm against speed cameras, I think they are an excellent idea in the right place. (Male, late 50s, professional driver focus group)

Well, the two cameras that piss me off, because I go to Manchester quite a lot, and I think it's the A55, two cameras and I know exactly where they are and everyone else knows exactly where they are, so we're belting along at 70 mph and this used to be a derestricted road, had no speed limit for many years, and then they resurfaced it and decided it was going to be a 60 on one part and then for some reason 50 towards the end, and you know the speed cameras are on straight stretches of road. Completely straight. And I can never work out, I mean fine on a Monday morning rush hour, you can't *go* faster than 40 mph along there anyway, but on a Sunday evening, no traffic around. What is the point? I mean what *is* the point? Other than generating income and trying to catch you out. (Male, late 50s, convicted driver focus group)

When drivers believe that limits are inappropriate representations of levels of risk, but hear the authorities defending them on this basis, then alternative rationales for this enforcement are able to exist. The perceived oversimplification brought about through the use of a simple-to-enforce schema of acceptable levels allows inferences to be drawn about the trustworthiness of the authorities. In some cases, the inappropriateness of travelling *at* the speed limit or even *below* it is noted, along with the authorities' apparent lack of interest in detecting this kind of dangerous behaviour, despite its demonstrable riskiness. Similarly, the ability to detect only one type of offence makes the speed camera a target for sceptical criticism. This was a regular theme of discussions at Speed Awareness Course sessions, summarised by one attendee: 'OK, so they've been rebranded as "Safety Cameras", but they don't protect us from drink drivers, unsafe parking, slow drivers, inconsiderate drivers, road rage, etc. Why only speed?' (female, early 50s, Speed Awareness Course).

Even some drivers who professed to be supportive of attempts to lower speeds generally were sceptical about the use of cameras to achieve road safety aims. They noted the potential danger of travelling at an inappropriate speed, something which, again, the speed camera could not detect:

Dick: A lot depends on whether they're driving that car to kill, deliberately, recklessly or whether they're just not aware of the speed limit.

Pat: But you can get killed without speeding, can't you?

Dick: But the campaigns suggest drive at 30 [mph] and you are safe.

Pat: And you're not necessarily.

Dick: What speed is safe? The speed is only as safe as the driver behind the wheel and the conditions he's driving in at that time. (Professional driver focus group)

If I use my own example [where his son had been involved in accident], is 30 mph 'safe'? Well, no, it isn't if you're passing a fag back to somebody, is it? But there is *too much about speed* ... Nothing to do with speed limits, that's about safety *within* speed limits. How do you slow vehicles down within the limit? (Male, early 60s, professional driver focus group)

This perspective was also evidenced in Blincoe et al.'s research, where the comment's author also took the opportunity to question the expertise supporting the use of speed cameras: 'The concept that "speed kills" put out by the police, government and others is a complete and utter fallacy that has no valid scientific proof. Careless and dangerous driving can kill, speed in itself does not' (male, 38, quoted in Blincoe et al. 2006, 375).

The authorities' claim to be motivated by road safety concerns is called into question through their lack of interest in other perceived risk-producing behaviour and their desire to punish behaviour that doesn't appear risky. As a result, ulterior motives that are neither fair nor just can be suspected, such as the ever-present allegation of revenue raising.[11] Procedurally just policies are considered to be those that are based on sound and legitimate reasons for their existence and which are operationalised in ways that appear consistent with these motives. Unfavourable outcomes are therefore partially legitimated by the rationale that justifies them. In the altered expert context already established, the formulation of accepted and persuasive causal chains underpinning the introduction of regulation of *mala prohibita* offences is particularly problematic and ulterior motives can be suspected.

Crucially, both the suggested alternative rationales for enforcement discussed below relate to the ways in which the authorities' stated interest in road safety apparently fails to be reflected in their chosen enforcement methods. If risk is not accurately reflected in policies that claim to reduce it, then alternative explanations for such policies are able to exist. Such alternative rationales fail, in various ways, to be procedurally just. This first alternative explanation for the prioritisation of speed limit enforcement in evidence in the debate centres around the professed

---

11 Qualitative research by Soole, Lennon and Watson (2008) in New South Wales, Australia, also supports this suggestion.

belief that the authorities (primarily national government) are enforcing speed limits because they are for some reason fundamentally or ideologically opposed to the car as a means of transport. As such, they are viewed as discriminating against a particular group by waging a 'war on the motorist'. Such a motive can be seen as reflecting Tyler and Lind's (1988) notion of bias on the part of the authorities. The stated aim of casualty reduction can thus be viewed as a smoke-screen allowing motorists to be targeted as a group by a government that is opposed to car use generally:

> I just view all cameras as part of a war on the motorist. Parking, fuel, speed, it's all the same. I saw a sign the other day that said 'pleasure drive' and that made me laugh. There's no pleasure in driving in this country, not now anyway. (Male, late 20s, experienced driver focus group)

> I've reached the conclusion that they are just anti-car. Every day it seems like a new bit of legislation comes out that in some way makes life harder for us. Either that or it's petrol costs that price us off the road. It's tempting to become a pedestrian, but maybe that's the idea! (Male, mid 50s, experienced driver focus group)

This alternative rationale is supported by media descriptions of motorists as the government's 'public enemy number one' (*The Guardian* 2004b; *Daily Telegraph* 2004), seen by some as 'a declaration of war on the 50 million men, women and children, [who] as drivers or passengers, are regular car users' (Rutherford 2001). The coalition government's announcement, in July 2010, that central funding for speed cameras would end was very explicitly linked to this sense of victimisation. Road Safety Minister, Mike Penning, described that development as 'another example of this Government delivering on its pledge to end the war on the motorist' (*The Independent* 2010b), a pledge that he had first made only days after taking office and which no doubt confirmed the existence of this 'war' under the previous administration in the minds of many drivers.

By constructing the authorities as biased and discriminatory, the policies they justify on the grounds of safety can therefore be dismissed as unfair. Whatever the reason for targeting motorists, it is not seen to be related to safety, and as such the regulation and restriction of motorists is viewed as unjustified victimisation. The extensive efforts of the authorities to convince the public that there are legitimate road safety reasons for the use of speed cameras are therefore greeted with scepticism. If road safety were the motive, some drivers suggest, the authorities would show an interest in other types of risky behaviour such as risky speeds below the speed limits, would enforce limits in all dangerous locations and would be prepared to excuse behaviour that (although technically illegal) caused no discernible risk. Instead, they are seen to operate an oversimplified decontextualised system that, while being seen to inaccurately represent 'real' risk, allows for drivers to be discriminated against.

A second, frequently heard objection to the use of speed cameras suggests that the policy is motivated by a desire to make money from speeding drivers – interpreted as blatant 'self-interest' on the part of the authorities (Tyler and Lind 1988, 163). The authorities' insistence on enforcing speed limits in locations that are thought to be safe and their refusal to acknowledge occasions on which offences seemingly posed no risk are seen as evidence of a desire to increase public funds rather than decrease road casualties.[12]

Of a sample of over 1,000 comments from internet-based forums, over a third mentioned this alternative explanation, while 82 per cent of national newspaper articles analysed did the same.[13] Major newspaper articles, particularly in the tabloid newspapers, have reinforced this ulterior motive, with headlines such as 'England's most lucrative speed camera revealed – raking in £2.3million in five years' (*Daily Mail* 2009), 'Traffic camera rakes in nearly £1 million in a month' (*Daily Telegraph* 2010a) and, explicitly linking location to perceived motivation, 'Outrage at speed camera that rakes in £1.3m a year on safe road' (*Daily Express* 2010a). Such headlines have been a consistent theme of the debate, spanning the period of the NSCP and beyond:

> FATSO GATSO: Speed cam nets £1m at roadworks. (*The Sun* 2006)

> Police profits before lives: Shocking new evidence of the way speed cameras are being used to milk motorists for cash was revealed last night. (*Daily Express* 2002)

> Milking the motorist: For some time Britain's middle classes have been feeling that they are the milch cows for this government. (*The Times* 2003)

Such has been the persistence of these claims that even the decision to cease funding for local authorities, explicitly described as being designed to discourage camera use, has been met with claims of revenue raising, evidenced by headlines such as 'Speed camera "cash cow" dries up' (*The Independent* 2010a). Whereas central funding was presumably judged to be motivated by the promise of revenue from fines from the cameras it encouraged authorities to use, the ending of this funding is described as profiteering because the government still receives the money from fines, but no longer has to invest its own money in order to do so (*Daily Telegraph* 2010c).

Given the perceived lack of convincing evidence of the effectiveness of the scheme, combined with the methods used to enforce limits, ulterior motives for

---

12   This suggestion was initially made because Safety Camera Partnerships were able to use the money that they collected in fines to fund further enforcement activity, and could therefore be seen to have a vested interest in catching as many motorists as possible.

13   Of a sample of 71 national newspaper articles, 58 mentioned a revenue generation motive, whether or not the article ultimately subscribed to that belief or not.

its existence can therefore thrive. Such ulterior motives do not, however, exist in isolation. Research shows that the operation of procedurally unjust policies that are perceived to be illegitimately motivated has consequences in the form of various types of withdrawal of consent from, and damage to the perceived legitimacy of, authority: 'If people feel unfairly treated when they deal with legal authorities, they then view the authorities as less legitimate and as a consequence obey the law less frequently in their everyday lives' (Tyler 1990, 108).

Responses of the type analysed here contain evidence of new and worsened relationships between some drivers and authority. Evidence from a variety of drivers suggests that their 'unfair' experience of the policy has resulted in a relationship now characterised for some by distrust and contempt, with individuals claiming to have 'suspended' their trust in authority (Giddens 1991, 142). This is a common theme of internet discussion forums,[14] and was strongly in evidence in both Corbett and Grayson's and Blincoe et al.'s research, which contained comments such as:

> When I got points for speeding at 2 a.m. I became resentful and at odds with authorities instead of working with them. (Driver quoted in Corbett and Grayson 2010, 6)

> Whoever has the job of placing cameras should be ****** shot. If the police want respect, start using some common sense. (Male, 60, quoted in Blincoe et al. 2006, 374)

Such a sense of 'alienation' (Corbett and Grayson 2010, 6), as well as 'anger' and 'lost confidence in the police's judgement' (Corbett 2008, 137), was, according to Blincoe et al., most commonly expressed by 'conformers' (one group identified by Corbett in her 2000 typology). They note that 'many felt that they were innocent, safe and experienced drivers who had been unfairly caught. This resentment manifested itself in lack of support for the cameras' (Blincoe et al. 2006, 377).

Corbett's 2008 article considering the 'current context and imminent dangers' of roads policing further connects the on-the-ground practices of roads policing to the broader issue of police legitimacy. She concludes:

> Future road law-enforcement policy and practice will inevitably be driven by technology, and potential problems associated with this cannot be ignored. To allow the imminent dangers ... to drift unresolved could augur badly for police in general, with disaffection or lowered confidence in the police among the mass of drivers leading to a public backlash culminating in an eventual challenge to the legitimacy of the police service itself. (Corbett 2008, 140–41)

---

14   See, for example, http://news.bbc.co.uk/1/hi/talking_point/3390665.stm.

Clearly the methods through which control is achieved, combined with the resulting ideas of victimisation, discrimination and revenue raising, are fundamental to the forming of these views, which may then be linked to a reduction in trust and support for the authorities who are implicated in this policy. Ultimately, therefore, given that the state believes that ensuring compliance with limits is the way to reduce casualties and thus to achieve the targets on which they have asked to be judged (DETR 2000a; DfT 2009b), the operation of policies that potentially reduce the levels of compliance and cooperation expected from the target population are potentially counterproductive. Tyler and Huo have noted, furthermore, that assessments of authority are 'not predominantly linked to performance-based judgments … Instead people's main consideration when evaluating the police and the courts is the treatment that they feel people received from those authorities' (2002, 196). This suggests not only that performance targets are not the measures by which authorities are judged by the public, but that attempts to achieve these targets through the use of automated methods may in fact contribute to the problem such interventions were designed to impact upon.

Conclusion

A preoccupation with risk has led to the reconceptualisation of behaviour into dichotomies of 'safe' and 'dangerous' and to the favouring of certain forms of detection and prosecution technologies. Such technologies appear to offer the potential for consistent and therefore (in some senses) fair enforcement. As important points of interaction between individuals and authority, however, they have also changed these interactions in ways that have deprioritised other antecedent elements of procedurally just practices. This analysis suggests that, rather than operating on simplistic notions of fairness in terms of consistency and impartiality, dissatisfied drivers are demonstrating that they require more contextualised and inconsistent treatment when it comes to systems that acknowledge their own individuality and the contexts in which they offend. Technologies that operate on a dichotomised interpretation of good and bad, risky and non-risky behaviour are seen as inadequate for the task of judging real-world, qualitative, contextualised events. The fact that such interpretations are conditional on human enforcement and on interpersonal encounters, and are therefore inapplicable to human–machine interactions, represents a challenge for future enforcement based around infrastructure of this nature.

No evidence of negative discrimination on the grounds of race, age or gender was noted at any point in this research and as such it might be suggested that the potential identified by Lianos and Douglas (2000) has been achieved in one respect. What this analysis shows, however, is that this negative discrimination has been eliminated along with a positive kind of discrimination that emerges as essential for a system to be considered procedurally just. The analysis undertaken here therefore suggests that the potential of enforcement technologies such as the speed camera, while great in terms of eliminating negative discrimination as part of its detection of risk, is limited by its simultaneous eradication of a separate kind of discrimination. The absence of discrimination welcomed by Lianos and Douglas is, therefore, not to be understood in any simple sense as a positive development brought about by the use of techno-fixes to control and minimise risk.

Procedural justice advocates have not, to date, considered the antecedents of procedural justice in anything other than *interpersonal* encounters between human controllers and the human-controlled. As such, the consistency they claim is required is always going to be mediated through those interpersonal encounters. It is not 'pure' consistency and non-discriminatory enforcement of the kind noted by Lianos and Douglas, but a kind that is produced in traditional, non-automated and hence *human* enforcement scenarios. The techno-fix, potentially the fairest form of enforcement to date, is therefore experienced as one of the most unfair forms, given that it throws out the baby of respect for relevant difference with the bath water of prejudice. Some alternative methods of enforcement which have the potential to be seen as procedurally just, and which were suggested by drivers themselves, form part of the discussions and recommendations set out in the next chapter.

Chapter 7
Developments – Past, Present and Future

This chapter considers the policy implications of some of the issues raised in this research, as well as some of the policy developments already introduced to modify and adapt the way in which speed cameras have been used since they first appeared on the UK's roads. The themes of expertise, identity and fairness continue to influence the analysis, which now focuses on the practical consequences of thinking about the speed camera debate from the theoretical perspective of risk. Having previously considered the nature of objections to speed cameras, and proposed some reasons for their form and intransigence, this chapter considers what lessons can be learned from exploring the debate in this way.

At the time of writing, one of the most fundamental challenges to the use of speed cameras to enforce speed limits is ongoing, with the withdrawal of central funding for the technology looking set to reduce the numbers of active cameras on UK roads.

While a central theme of this research has been that there are numerous important and valid reasons for taking drivers' concerns seriously, this does not necessarily mean that the entire policy should be abandoned, nor that there are major public opinion points to be scored by doing this. If properly acknowledged and understood, the concerns of some drivers can be put to constructive use. This chapter therefore considers ways in which the considerable speed camera infrastructure that has emerged since 2001 can continue to be used, but in ways that reinforce rather than undermine the legitimacy of the authorities while maintaining their effectiveness at securing prosecutions that contribute to road safety objectives. The suggestions presented here therefore have an eye towards both acceptability and effectiveness, given that compliance is linked to both and given that compliance continues to be a central aim of road safety strategies.

This chapter should not be viewed simply as an instruction manual for making speed cameras more effective at catching drivers who break speed limits. Its purpose is not to make it easier for government to operate technologies of control in any simple sense, and its message is that technologically successful interventions may still fail if their operation fails to take account of the populations on whom they operate. The purpose of this chapter is to recommend the best ways in which the technology can be used to secure reductions in death and injury where it is the most appropriate intervention, but in ways that do not undermine the confidence, consent and cooperation shown to the authorities implicated in their use. Operating procedurally just enforcement polices is crucial to securing the consent on which policing in the UK relies, but it should not be used to cover up greater injustices in the system or as a method of 'manipulation and exploitation' by the authorities to encourage recipients to accept otherwise unjust policies (MacCoun 2005, 193).

Although this study has deliberately concentrated on oppositional voices in respect of speed camera use, it is clear that there are many who are strongly in favour of the use of speed cameras and who feel protected by their presence. As such, although the withdrawal of funds from local authorities (at least partly intended to force a reconsideration of the way in which road safety is pursued) by the coalition government in 2010 looks set to move the emphasis away from speed cameras, the wholesale abandonment of the use of speed cameras looks both unlikely and inadvisable. Corbett (2008) notes that the future of road safety policy and roads policing will inevitably continue to involve expansion of the use of technology, and this chapter also considers the lessons from this debate for other technological interventions such as ANPR (Automatic Number Plate Recognition), before broadening the focus further to consider other enforcement contexts with a similar profile.

Responding to the Challenges of Expertise

Hebenton and Seddon argue that the democratisation of expertise poses challenges to 'decision makers', but that the phenomenon should not necessarily be suppressed. They argue that the extension of the notion of expertise can be achieved through the recognition of 'various forms of "knowledge" in decision making itself' (2009, 357), implying that such expansion of what it means to be an expert should be viewed positively. This research, however, demonstrates some of the challenges that potentially result from a situation in which those public, experiential knowledges come to exert influence over more traditional experts, and which contribute to the construction of stubborn positions of resistance among populations that are simultaneously responsible for risk and at risk of other risky behaviours. Regardless of its positive or negative impact on policy, such a reality needs to be acknowledged and interventions designed with an eye to both their effectiveness *and* acceptability. The arguments set out in Chapter 6 indicate why these two aspects may in fact be mutually inclusive, given that compliance (as a consequence of acceptability) is central to the effectiveness of interventions that rely on individuals changing their behaviour in order to reduce the occurrence of activities identified as risky. One way in which individuals can be persuaded to change their behaviour is by securing their normative compliance to the laws regulating that behaviour. Achieving normative compliance is a less resource-intensive and less onerous way of generating compliance than instrumental compliance. It, too, is dependent on the authorities' treatment of the public:

> If the effectiveness of legal authorities ultimately depends on voluntary acceptance of their actions, then authorities are placed in the position of balancing public support against the effective regulation of public behavior. Legal authorities of

course recognize their partial dependence on public goodwill, and are concerned with making allocations and resolving conflicts in a way that will both maximize compliance with the decision at hand and minimize citizens' hostility toward the authorities and institutions making the decision. (Tyler 1990, 24–5)

Consequently, attempts to convince the public that the policy is legitimately motivated and operationalised are vitally important. Given the context identified in Chapters 3 and 4, however, this is also an increasingly difficult mission. This section considers ways in which the authorities have attempted to emphasise the legitimate basis on which a speed camera is installed and ways in which this can be taken forward.

Publicity and Promotion

The relationship between the traditional experts in the speed enforcement debate and the general public has already been identified as a source of some of the problems encountered. Traditional experts, both in the government and Partnerships and within the various pressure groups, assume naivety on the part of their public audience, a misunderstanding that causes them to assume that the ignorance of the public can be overcome by the supplying of more and more detailed and 'conclusive' information. As has been shown, the context of the debate and the focus on issues about which the public may perceive that they have a level of expertise of their own mean that public opinion cannot simply be won over by impressive statistics and independent polls. However, the general motoring public has not usually been considered worthy of consultation on the issues within the debate and its objections and observations are more often than not dismissed as either the bitter complaints of 'the motoring lobby' or of 'speed freaks'. The perceived need to deny, publicly, that a debate was even occurring has meant that, for much of the last decade, criticisms have been both denied and ignored.

However, in response to the (privately at least) acknowledged debate, attempts have been made to amend the policy to render it more acceptable. But because they have not generally listened to the substance of the criticisms (and even, occasionally, compliments) of drivers, the authorities have made amendments that they have presumed will improve the acceptability of their policies, while not really being aware of what makes them unacceptable in the first place.

Efforts made to date have stressed the way that the policy meets the criteria the authorities themselves have set with regard to openness and measured effectiveness (for example, DfT 2004d), and have left them wondering why the public remains unmoved, when the public is operating on an entirely different set of criteria. Johnston suggests that, within policing, accountability has been replaced by concepts such as 'balanced budgets' which become proxies for genuine accountability (2000, 32). Likewise, the complex issue of 'fairness' has been replaced by the more measurable concept of 'openness'. Numbers of signs, the publication of crash data and the release of the audits of Partnerships can all be given as examples of the openness and transparency of the authorities, but will not overcome public doubts if they fail to address (and therefore simply publicise and draw attention to) issues of unfairness and illegitimacy.

Early publicity of the camera scheme made the mistake of giving numbers of drivers caught at different locations as a measure of the scheme's success. This indicator is seldom, if ever, used now, suggesting that the authorities have gained some awareness of the judgement criteria used by the public – and that this approach does not meet them. The scheme, and its targeting of 'respectable' people and 'normal' drivers, is not conducive to demonstrations of its effectiveness based on measures used, for example, to indicate a force's success at catching burglars. Instead, members of this target group assess their authorities on the fairness and justness of their encounters with it, as offenders.

Further, by publicising the scheme's 'success', measured in terms of drivers prosecuted, the authorities would simply have been drawing attention to the fact that speeding is a majority activity, and therefore one entirely consistent with such majority identities as 'respectable' and 'law-abiding'. Doing so draws drivers' attention to the fact that their behaviour is not, statistically at least, deviant. It also serves to reinforce any sense of injustice by reminding drivers just how many 'innocent motorists' have been punished and how much revenue has been generated. This, again, links back to compliance issues, given that the perception that you are the only one consenting is enough to bring about non-compliance, even if policy makers successfully reinvent themselves as trustworthy (Levi 1997). According to these observations, the promotion of *falling* numbers of both speeding drivers and of speeding prosecutions may be a more successful tactic in generating the belief that the activity is becoming more marginal and deviant and less socially acceptable.

Conspicuity and Publicity of Camera Locations

The signposting of speed cameras and stretches of road where enforcement may take place was made part of legislation following recommendations made in the Police Research Group study (Hooke et al. 1996, 41 and 43–4), with strict requirements as to the amount, type and style of signage introduced at the same time as rules about the conspicuity of cameras (DETR 2001). These rules applied to any cameras installed under the NSCP for which Partnerships wished to reclaim costs. Following the ending of the NSCP the rules are no longer compulsory.

Safety Camera Partnerships now publish the locations of their fixed speed cameras on their websites, and announce where and when mobile speed cameras will be operating via local radio and newspapers. In addition, many road atlases now feature all fixed camera locations. However, this was initially an 'underground' activity carried out by opponents of speed cameras, who invited visitors to their websites to advise them of locations in an effort to better inform drivers and essentially to 'try to outwit the Partnerships' (interview with creator of anti-speed camera website).

Rather than welcomed by drivers as concessions by the authorities, the publication of camera locations and their signing seems to be viewed as no more than drivers deserve. There appears to be an expectation that drivers are entitled to be told where and when their illegal behaviour is actually likely to be considered a problem by enforcers.

During the pilot phase of the NSCP, speed camera housings were grey in colour. This was not, according to the Home Office, a conscious choice by the authorities, but an artefact of the manufacturing process (interview with senior Home Office official). In December 2001 a change in the law required that they be changed to a high-visibility yellow. The change was, according to one Home Office official, motivated largely by the desire to be seen to be treating drivers more fairly, and to remove the prospect of drivers accusing the government of trapping them or using underhanded methods:

> Not hiding them; that was a main point, that everyone can see where they are and then if you don't see them, then it's your problem if you get a speeding ticket. But we did get letters from the other side saying, 'Why are you making them yellow? It's too easy now.' There was a legal challenge which was lost. The pro-road-safety groups will always say that they should've been kept grey. (Interview with senior Home Office official 2004)

Official statements confirmed this concession, taking the opportunity to deny claims of 'revenue raising':

> These rules should ensure that motorists are not caught by surprise by cameras. I hope that this will reinforce the Government's message that cameras are there to save lives at places where there is a history of speed-related accidents. They are not there as a means of raising money. (John Spellar, Transport Minister, quoted in DETR 2001)

This development was one of the earliest concessions to motorists, who felt grey cameras were 'traps' and a method of covert enforcement. The then government acknowledged that 'no firm evidence exists one way or the other on whether or not highly visible cameras obtain better casualty reduction than less visible or hidden cameras', but 'numerous complaints from drivers have been received that cameras have not been evident'.[1] The actual experience of interaction with the technology

---

1   This quotation was previously (in 2003) included as part of the Frequently Asked Questions section of the speed camera section of the DfT website. Although it no longer appears in this document, it remains in evidence on several Partnership websites, including   www.cumbriasafetycameras.org/press/Safety%20Cameras%20FAQ.pdf   and

at that moment of passing it is made more 'fair', but possibly at the expense of road safety objectives.

As with most developments, this change met with mixed views from the participants in the debate. The change in colour of camera housings was, as mentioned by the Home Office official above, accompanied by a legal challenge in 2002 from two road safety groups[2] who argued that the change in law made cameras too obvious, leading to the possibility that 'speeding drivers will brake before a camera then speed away after it' (Transport2000 2002). Their fears seem to have been confirmed by the high proportions of drivers deemed 'manipulators' in Corbett and Simon's (1999) typology that are in evidence in research into drivers' responses to speed cameras (Blincoe et al. 2006). Some academics also opposed the change on the grounds that it 'risk[ed] defeating the object of installing the cameras in the first place' (Keenan 2002, 154). The issue demonstrates the competing priorities faced by government, and the fact that public opinion is clearly one of these concerns. The issue concerns the question of whether making cameras visible encourages drivers to believe they can speed with impunity everywhere else, or whether it generates greater compliance via increases in perceived legitimacy because drivers are not caught unaware. Clearly, the 'yellowing' of cameras makes it more likely that drivers will comply with limits *at speed camera sites*, where the statistics which demonstrate the success or failure of the scheme are generated. On the reverse side, grey cameras would either generate increased compliance across the whole network (by virtue of potentially being *anywhere*), or result in the detection of more offences (as drivers fail to notice them and slow down). The authorities defended the decision on the grounds that cameras were not the only enforcement methodology being used and drivers would still be compelled to drive at or below the speed limit in all areas: 'Part of the reason for the challenge [by the two road safety groups mentioned above] failing was the understanding that the police could and did enforce speed limits away from camera sites using other hand held equipment' (DfT 2005a, 23). It was clearly still assumed that drivers would not be speeding at other places *just in case* other methods of enforcement were in use.

The change seems to have been a compromise measure, driven at least partly by the 'huge public backlash against the apparent "profiteering" of the local authority at the expense of the "unfairly-trapped" motorist' (Keenan 2002, 154). There seems to be little chance of, or demand for, a U-turn on this issue, and as such it should be assumed that cameras (where they are still used) will remain yellow in the future, potentially being made *more* visible rather than less.

The issues raised in this context are about whether the authorities are attempting to deter or to catch and prosecute motorists. Again, this encourages drivers to think only of risk in terms of the risk of detection and not in terms of risk to

---

http://microsites.lincolnshire.gov.uk/lrsp/section.asp?catId=20496 (retrieved 23 August 2010).

2   Transport2000 and the Slower Speeds Initiative.

others. Locations that are not publicised as having an enforcement presence are, by inference, more 'safe' than those that are, either because they have not been considered as worthy of enforcement as other locations (less of a crash risk) or illegal behaviour is announced as being unlikely to be problematised by the state at these locations (less risk of enforcement). The spectre of additional enforcement in the form of police patrols is therefore crucial to maintaining legal speeds across the whole network, not just at camera sites, and any further reduction in the level of police patrols risks undermining this. Corbett has identified that many drivers perceive the risk of detection by a traffic officer to be low, and consequently feel that they are 'safe' speeding other than at camera sites (2008, 135).

Detailed Explanation of Crash Risk

Although the publicity of camera locations is now an accepted part of the policy, this amendment seemingly confuses the issue of *openness* with that of *fairness*. Cameras can be publicised widely and openly, but if they are not perceived to be legitimately located, then this strategy only publicises something that can cause damage to the perception of the policy and the legitimacy of the authorities behind it. In addition to signposting the locations of cameras, drivers themselves suggested that more specific information about the nature of the risk in that location would be appreciated. Current policy is to use a sign bearing a camera symbol to indicate stretches of road where enforcement takes place. Drivers may indeed make the connection between the sign and the criteria for placing cameras, thus reaching the conclusion that they are passing an area where there have been recent casualties, but one Speed Awareness Course explicitly encouraged drivers to think of each camera as representative of bunches of flowers (as often laid at the site of fatal road accidents by those affected). The conveners of the course drew attention to the 'KSI' rules, suggesting that 'each camera represents at least three deaths'.[3] Drivers did express support for making the background story of each camera more explicit:

> There's only one camera I've ever seen which actually said 'This is an accident black-spot', yeah? I mean I don't like the cameras at all but they do slow traffic down. I think there are other ways of slowing traffic down if they were dedicated to actually doing that, but one of the ways is highlighting the dangers of certain parts of the road system and there have been lots of accidents. Yeah, so if you put the *reasons*. (Male, late 50s, convicted driver focus group)

> *Pete*: What they should do is promote how many deaths or accidents or whatever happened in that area and then people might see the point, because they're always getting bad press reports, aren't they? I mean, I've read a press report that says speed cameras haven't reduced a single road death in Britain. Now

---

3   KSI stands for 'killed and seriously injured' not only 'killed'.

whether that's true or not I don't actually know, but it's what I've seen in the paper.

Tom: But I don't think anyone can tell that, can they?

Pete: No, but it's in the papers, I've read it.

Tom: The one I was on about has definitely helped, it's definitely helped.

Pete: That's what I'm saying. You see it in a different light so if they *promoted* the reasons to people, people would understand, see it in a different light, wouldn't they? (Professional driver focus group)

Rather than abstract statistical findings that relate to large-scale and generalised research studies (of the sort the authorities have produced to date), these drivers are calling for the reasoning to be justified on an individual level. Suggestions were made that specifics of recent deaths or injuries would have more impact and also support the enforcers' claims that the camera was in a dangerous location. Such specifics could include the numbers of people killed and injured, the qualifying time period and the conditions in which the crash occurred. The addition of whether the victim was a car occupant or a pedestrian could also remind drivers of their responsibilities to other road users and encourage them to think of themselves as a risk-producing as well as an at-risk agent. The explicit stating of the reasons would, it is suggested, help persuade people that their actions may be genuinely risky and prevent them from excusing their behaviour quite so easily.

This localised information approach has been supported by some motorists' groups. In commenting on the potential for roadside 'shrines' to educate and prevent future accidents, the ABD note that, without specific details of crashes, 'it's all too easy for shrines to be used to fuel ignorance of the true causes of road crashes and to stoke up hysteria in favour of inappropriate and ineffective "safety" measures' (ABD 2004f). The examples given by this interest group do not include an example in which a crash was actually caused by excessive speed. Given the history of the group responsible for the suggestions, this is presumably meant to suggest that, by calling for this kind of information, the authorities would be obliged to reveal how few, if any, of the crashes being used to justify the placing of a camera were related to speeds in excess of the limit. From the opposite perspective, therefore, the providing of this type of explanation and information could be an opportunity for the relevant authorities to demonstrate the extent of speeding's involvement in crash causation. The 'real' contribution of excessive speed to the causation of crashes once again surfaces as central to the debate, central as it is to the legitimacy of a technology designed to reduce speeds to within legal limits.

Other research has similarly found that the location of cameras was a 'contentious topic', crucial to whether they were perceived as legitimate road safety tools or as tools for generating income. One driver made this point explicitly by noting: 'Cameras are about profits. If they were concerned with casualties or accidents they would be installed on housing estates' (Corbett and Grayson 2010, 5). The authors note that these views betray an ignorance of the siting rules that govern where cameras can be installed, suggesting that '[i]f the public are to have greater confidence and a better understanding of decisions on camera location, then more information about and explanation of the procedures, limitations, and rules involved could help to alleviate any perceptions of capriciousness or unfairness' (ibid., 4). This counters the 'commonsense' expertise of drivers with statistics, but the suggestions of publicising the *specific* KSI statistics that have justified each specific camera takes this idea further and, again, brings the large-scale, general policy context down to a local level. Drivers' anecdotal evidence and assumptions about where the dangerous locations really are may not be accurate, but unless this inaccuracy is challenged by information specific to each potentially delegitimising context it will continue to undermine the use of speed cameras that appear to be on 'safe' roads and thus appear to be placed for reasons other than road safety.

However, given the wider debate context in which the speed camera policy is located, it is understandable that large-scale findings are easily dismissed by drivers who have a vested interest in defending their own respectable self-identity and driving licence. Further, and given what is known about drivers' beliefs in their own superior skills, it is by no means certain that evidence of another driver's misfortune would cause most drivers to modify their behaviour. In this sense, therefore, publicity of the details of crashes at specific locations may not always achieve reductions in risk. However, it may help to legitimise the installation of specific cameras, particularly ones that are perceived to be on 'safe' roads. In this way, risk reduction may be achieved indirectly via the increased compliance seen to result from an increase in perceived legitimacy.

In its 2011 *Strategic Framework for Road Safety*, the coalition government has acknowledged the importance of information on-road safety interventions and activities as part of its dedication to 'empowering local citizens' (DfT 2011, 8). The framework advises that central government will be:

> [s]upporting the provision of local information to the public to increase scope for challenge by showing the level of risk geographically, the comparative road safety performance of different areas and service providers for different groups and information on all safety cameras. (Ibid., 9)

While it is not clear what type of 'information' is being offered, the increased accountability of individual cameras does seem to form part of the intention.

Addressing the Issues of Identity and Justice

Having considered the ways in which speed limit enforcement policy can respond, and in some cases has responded, to the challenges posed by the changed expert context in which it operates, this section moves on to consider the other challenges identified in previous chapters. The issues of respectable identities, and their enthusiastic defence, and of concerns about procedural (in)justice are taken together here. The proposed changes (some of which were put forward by drivers themselves) relate in large part to the reintroduction of *context* into the decision-making and enforcement processes accompanying speed cameras. This is achieved, first, through an exploration of issues including a more acceptable method of determining appropriate speed, which, it appears, drivers feel more accurately reflects the ever-changing circumstances in which acts of driving occur. Subsequently, two proposals for punishment methods that are seen as more fair than a fixed penalty are considered, before the analysis goes on to consider more fundamental criticisms of the use of the speed camera as the detection technology. This is followed by consideration of two alternative technological devices which, while sharing some features with speed cameras, were proposed as being potentially more acceptable.

Each section considers a more procedurally just alternative to the problematic elements of the system identified by drivers previously. Where prosecution systems based around strict liability principles and fixed penalty procedures have been favoured as a result of the centrality of the techno-fix to the policy, alternative systems that are seen to reflect the level of risk posed are considered, along with ways in which mitigation, context and voice can be respected. Further, while the speed camera has been the favoured techno-fix to date, the human traffic officer is remembered by some drivers as the fairer and more legitimate alternative detection method.

'Appropriate Speeds', Not Speed Limits

As an alternative to the oversimplification of circumstances that speed limits were often criticised for representing, many drivers suggested the notion of 'appropriate speeds', with the 'appropriateness' of any chosen speed dependent on its context. This notion of a contextualised approach is strongly in evidence in drivers' criticisms of the inflexibility of speed limits, inflexibility which they know from experience to be a poor reflection of 'real' levels of risk caused by drivers of different abilities in different conditions. Both 'inappropriate' and 'appropriate' speeds imply an element of choice, of free will and of interpretation, whereby speed choice reflects on-the-spot, contextualised judgements of 'actual' levels of risk given the ever-changing conditions and circumstances in which any act of driving occurs. By appealing to be allowed to use such interpretative skills, drivers and those who campaign on their behalf are asking to be given credit for their driving skills and their human intelligence (skills that are, in most cases, considered to be above average, of course).

It is these skills, and the desire that they be acknowledged by the enforcing authorities, that seemingly underpin speeders' claims that 'I wouldn't do it if it were dangerous' (male, late 40s, Speed Awareness Course) and 'I only speed when it's safe' (male, late 20s, experienced driver focus group). Although important and influential to these drivers, as measures of their agency and their moral autonomy, and acknowledgements of their skill, they are obsolete judgements given the regulatory framework of speed limits. Such a framework seemingly encourages them not to exercise their driving skills, but to 'drive by numbers' (ABD 2008) and allow limits, not their individual skills and abilities as drivers, to determine their speed choice. Such a reflexive, adaptive and, ultimately, *appropriate* framework is clearly heralded as the morally and contextually superior alternative to 'slavish adherence to a blanket speed limit' (ABD 1999d) and is crucially seen as the route to a more realistic, safe and procedurally just system. Instead of allowing this agency, many drivers support the view expressed in the motorcycling magazine *StreetBiker*, where it was felt that the authorities had 'declared war on all of us irresponsible enough to feel that we can ever "interpret" speed limits for ourselves' (Mutch 2002). Similarly, the ABD has chosen the slogan 'for drivers who can THINK for themselves', with the capitalisation of the word THINK seemingly a reference to the THINK! road safety campaign. Viewed again from a risk perspective, the ability to make choices for oneself and not be subject to either imposed risk or imposed safety is crucial (Adams 2003), with risk taking and independent risk management viewed as vital expressions of moral autonomy (Ericson and Doyle 2003, 17). According to these authors, therefore, attempts to deprive individuals of their rights to choose their level of exposure to risk will be viewed as attempts to deprive them of their freedom, autonomy and even elements of their identity, and are consequently badly received (ibid., 20).

Stradling et al., however, note that in their research '[t]he libertarian position, that each, on the basis of their own experience as a driver, should judge for themselves, from moment to moment, the speed at which it is appropriate to proceed, received little support' in their survey (2003, 110). I would suggest, however, that the question of whether drivers are considering their own driving or that of others is relevant here. Given the foregoing analysis, we might expect some reluctance to support the idea that other (presumably below-average) drivers be allowed to use their discretion, while supporting the idea when applied to our own driving. Whether or not drivers are making this distinction in Stradling et al.'s research cannot be determined from the presentation of the data in that study. The question itself appears to be phrased in such a way that invites drivers to consider other drivers rather than their own case, which supports the suggestion made here.[4] The qualitative, self-report study carried out by Blincoe et al. supports

---

4   Respondents 'agreed' or 'disagreed' with the statement 'Speed limits don't mean much on roads and drivers should judge whether to drive faster or slower'. The suggestion

this interpretation, as 'manipulators' (to use Corbett and Simon's 1999 typology of drivers) were found to express the view that 'limits and restrictions are useful for less skilled drivers' (2006, 376).

This issue presents some considerable challenges for enforcement based around the techno-fix. Enforcement in pre-automated times could be seen as being more 'accurate' in its targets, given that it resulted in the prosecution of 'bad' drivers – those drivers who were involved in crashes and who had therefore demonstrated their inferior skills. However, the new focus on risk as potential harm located in the future, and therefore on *preventing* crashes, means that this no longer occurs. Drivers caught speeding may be those who would never have caused injury, or they may be those who were, literally, headed on a collision course. It is not possible to identify which drivers had which destiny, nor is this desirable within the current policy, because the intervention is designed to reduce risk across the driving population and the road network by minimising potentially risk-producing behaviour en masse, rather than relating to specific individual events. Because many drivers believe that they are better than the average driver, such a blanket approach is likely to be experienced as unjust and unwarranted.

Reconciling demands for more reflexive speed choices with current enforcement methods is impractical. However, Blincoe et al. also found that the drivers in their research supported the idea of varied limits that would reflect weather and road conditions, for example (2006, 375). Such limits are theoretically enforceable using speed cameras and indeed variable limits are applied to some stretches of the UK motorway network, while France, for example, maintains two motorway limits, one for wet and one for dry conditions. However, such techno-fixed reflexivity is unlikely to appeal to all drivers, particularly those who believe that their skills exceed those of their fellow road users, whatever the conditions.

Graduated Points

Rather than consistently, neutrally and impartially dispensing the same penalty in all circumstances, many drivers preferred systems of punishment that involved some potential for the circumstances of an offence to be reflected in the penalty, rather than for each offence to attract the same fine and number of penalty points:

> *Terry:* If you're going to be punitive about it, then [the penalty] should reflect the severity of the offence, so, yeah, I think it depends on the speed you are going over the limit.

---

being made here is that a different answer could potentially have been given if the statement had been phrased in the first person: 'Speed limits don't mean much on roads and I should judge whether to drive faster or slower'.

Sue: We're all in favour that for doing 38 [mph] I should only have had one [penalty point], shouldn't I? Certainly not *three* for *eight* miles an hour! I think a sliding scale of points would be better.

Terry: Well, it's certainly more logical …

Chris: How about one point for maybe just a small infringement, ten points for a bigger infringement? They need to set up the law round the technology they've got rather than just introduce the technology and 'let's get 'em'. Then that would be a lot fairer, wouldn't it? (Convicted driver focus group)

I heard about this [consultation on the issue of graduated points] and I thought it sounded sensible. I thought, because then you're actually giving a punishment that more accurately reflects the offence. People who've just tipped over … you can see that they might be doing it by accident, but if you're doing a lot over, then it doesn't matter even if your speedometer's not working, you know that you're going faster than 30 [mph] anyway. (Male, late 50s, professional driver focus group)

Stradling et al. have also noted widespread support for varied penalties in their study, as '9 in 10 respondents (90 per cent) thought the penalty for speeding should vary with the amount by which the speed limit was exceeded' (2003, 110). Blincoe et al. quote drivers who felt that first-time offenders should be treated differently to other drivers (2006, 376), and a study by Autonational Rescue (a breakdown/ rescue service) also found that motorists believed that drivers who were caught speeding should be treated differently according to their circumstances, with speeding by unlicensed drivers and those exceeding the limit by more than 20 mph given increased penalties. For 'experienced drivers', those with a 'previously clean record', and those 'travelling just a fraction over the limit', however, there was support for reduced fines. The author noted that '[m]any people responded well to the notion of varying levels of speeding fines depending on the background of the driver and/or what speed they are caught doing. It's an interesting approach to a modern problem' (Autonational Rescue 2004).

The issue of graduated points has been under discussion in DfT publications for some time (see DfT 2004b, 2004c). Although the 2006 Road Safety Act made provision for offering a graduated response to speeding along these lines, with more penalty points being made available for more extreme offences, this option has not yet been taken up 'on the ground'.

The Transport Secretary at the time of launching this consultation noted that these proposals were for 'a fairer system of penalties for motorists caught speeding', conceding that the proposals would be 'more effective and appropriate than the current "one-size-fits-all" approach' (DfT 2004c). That the controversy around the NSCP played a key part in bringing about the changes is clear from the following quotation from a senior DfT official speaking in 2004 when the plans were at consultation stage:

> The original proposal was to look for a more serious penalty. Two years ago we were saying speeding is very dangerous, very severe, and anyone who does too much of it above a certain threshold gets an even stiffer penalty. So we consulted on that. What's now being proposed is, because the government has been perceived as being so nasty to motorists, at the lower end, the minor infringements, they want to be a little bit more proportionate. Offer a slightly softer penalty so that when you didn't *mean* to do it on this occasion, we'll just give you a minor reprimand, two points and only a £40 fine, and hopefully they won't do it again. So it's being a bit softer. (Interview with senior DfT official)

The reception from road safety groups was not positive, but must be balanced with the predicted gains in public support, as the following quotation demonstrates:

> They [road safety groups] don't think it's a good idea at all because it may convey to drivers that speeding is no longer regarded so seriously. There's a very confusing message. But if the message is such that drivers think they are being more fairly treated, and they're willing to accept that there is a bit of proportionality here and that cameras aren't black and white – I mean black and white in the judgement sense – and if they're only a few miles over, they get a lower ticket, then fine. But we shall see. (Interview with senior DfT official 2004)

However, in its 2008 consultation on the topic of road safety compliance specifically, the proposals had been refined so that they only included an 'upward graduation' – an increase to six points for the most extreme speeding offences. It was intended that '[t]his will effectively target excessive speeders, and move them more quickly to the 6–9 point threshold where the evidence shows they will slow down' (DfT 2008, 27, referring to research by Corbett et al. 2008). The option of a 'lower graduation' was also considered but was dismissed as undesirable given that ACPO guidelines recommend a certain level of tolerance before enforcement occurs in any case and that any detected speeder must therefore have been exceeding the speed limit by a significant amount to have been caught (DfT 2008, 28).

As well as failing to offer 'softer' punishments for those at the 'lower end', this approach to graduation does not offer the potential for the 'softer' contextual issues raised by drivers such as the 'respectability' or 'skilfulness' of the accused, the 'actual' level of risk or intent or lack of it to be acknowledged. Intent or lack of it is, seemingly, represented by the assumption that a minor infringement may be accidental but that a major one is intentional. 'Time of offence' and 'degree of excess' are therefore potentially new proxies for level of risk posed and driver culpability, meaning that there are new proxies for 'safe' and 'dangerous', 'intentional' and 'unintentional' and potentially (by extension) 'bad' and 'good' drivers, 'bad' and 'good' people.

While potentially having some appeal, this kind of 'techno-fixed proportionality' would not acknowledge the more subjective aspects of context, such as drivers' 'respectability', inherent 'law-abiding' self-identity, or the fact of their previously 'unblemished driving history' – considerations demanded by many drivers and considered previously. Similarly, the geography of offences (identified as important by numerous drivers) is not represented, for instance, by the introduction of higher penalties for offences near schools and lesser penalties for those on motorways and/or in the middle of the night.

To phrase this in procedural justice terms, it seems that, rather than consistently, neutrally and impartially dispensing the same penalty in all circumstances, preferred systems of punishment seemingly involve some potential for the circumstances of an offence to be reflected in the penalty, rather than for each offence to attract the same fine and number of penalty points.

As well as failing to be sufficiently flexible, the type of graduated penalties envisaged by the DfT would still be pursued within the existing fixed penalty structure, removed from the courtroom and continuing to deny the offender voice and respect. The many and various contextual aspects deemed important by the accused driver would still fail to be heard, with the slight increase in flexibility in the system still confined to *what drivers do* and failing to respect *who they are*. For this reason, the introduction of graduated points, should it occur, should not be expected to silence drivers' criticisms that relate to the acknowledgement of context in speeding prosecutions. In dismissing the case for downward graduation, the DfT's 2008 consultation did draw attention to potential areas of discretion that were envisaged as helping avoid the criminalisation of the least deserving drivers. They noted that: '[t]he police will continue to have discretion over when they penalise drivers, and the option to give a warning or offer a speed awareness course in lower-level cases' (DfT 2008, 28). The role of police discretion and the issue of driving training alternatives raised in this quotation are considered in the next two sections.

The Speed Awareness Course (SAC)

Speed Awareness Courses are alternatives to prosecution which normally involve education and some on-road tuition. The authorities' motivation for allowing fine revenue to be spent on educational courses is summarised by the DfT official below:

> [T]he motoring groups' oft-expressed view is that instead of just prosecuting drivers you should be advising, informing, educating them, which is fair enough. So if you go to a Speed Awareness Course and the reason that speeding is dangerous is explained to you in a rational way, then in future you might think, 'Yeah, I'll slow down for my own good and the good of others'. It's beneficial

> to people to understand why there is enforcement, and maybe improve not just their behaviour, but their *attitudes*. Because you can change behaviour with cameras, but you can't change attitudes and the two are quite different ... It's what we've done with drink driving. (Interview with senior DfT official)

As such, the aim of generating normative compliance with the law is clearly a motivating factor, with normatively compliant motorists therefore kept away from the enforcement system and any potentially delegitimising consequences it may have. The influence of other voices in the debate is also clearly a motivating factor.

The method is considered to offer an opportunity for the exercise of discretion by the police, who can decide on the most appropriate way to deal with motorists after they have been detected by a speed camera or police officer (ACPO 2009). As well as education, courses offer the opportunity for drivers to avoid paying the usual £60 fine and incurring the usual three penalty points (with drivers paying the price of the course instead). The opportunity of attending a course is generally offered to first-time offenders who meet locally set eligibility criteria in terms of their recorded speed and who have offended in any limit other than a 20 mph limit. Potential attendees of courses are therefore selected on the basis of their risk profile, with only drivers doing speeds within a certain threshold being considered likely to benefit from the education the course offers. Drivers are thus allocated to interventions 'by *sorting* (for risk) rather than by *tailoring* (for either culpability or need)' (Sparks 2000, 132).

Only one opportunity is generally offered in a three-year period, so subsequent detections within that period would result in prosecution (ACPO 2009, 5). Courses are generally run by local Safety Camera Partnerships and may involve outside contractors such as advanced driving instructors. The driver can accept the offer of attending the course or decline it, in which case they are prosecuted through the usual fixed penalty system. Some areas operate courses that are entirely classroom-based, while others include a practical, on-road element. By teaching skills for identifying speed limits, courses can attempt to address people's complaints that they offended because they were unaware of the limit. By trying to prevent such potentially indignant offenders from entering the system, the courses may diffuse some criticism.

Courses of this nature emerged in a piecemeal fashion across the UK, until 2009 when ACPO guidance intended to contribute to a standardisation of the format was issued (ACPO 2009). Some leeway exists for Partnerships to design their own content and, as noted above, some courses contain both practical and theoretical elements while others are purely classroom-based. This could potentially raise issues of consistency across the country with different costs and access requirements operating.

In terms of their effectiveness at securing attitude change, evaluations have found moderate effects. McKenna compared attitude change between the start and end of a course for 6,401 drivers attending SACs at eight locations and concluded that:

The speed awareness workshop was shown to produce small to medium differences in drivers' attitudes to speeding, their perceived social pressure against speeding and their perceptions that they could control their speeding in the future. The workshop diminished the belief that speeding is enjoyable and increased the perceived legitimacy of speed control. At the end of the workshop, drivers were more than four times more likely to disagree that driving at 35 mph in a 30 mph limit is safe. There were also clear differences in speeding intentions. For example, drivers at the end of the workshop were more than five times more likely to intend to keep to the 30 mph limit. (McKenna 2007, 274)

Other research by McKenna and Poulter found that participants showed more positive attitudes towards speed controls than those disposed of via other means and that these effects persisted at six months. Participants' self-reported speeding in 70 mph limits was unaffected, but their compliance with 30 mph limits had increased (2008, 4).

However, effectiveness aside, it might be argued that an equally important issue to be addressed by SACs is the provision of the opportunity for drivers to have 'voice' within the system, and this opportunity is offered to all drivers by virtue of their presence in a room with a human representative of the policy. Tyler and Lind note that '[t]he perception that one has had an opportunity to express oneself and to have one's views considered by someone in power plays a critical role in fairness judgements' (1988, 106). Although such courses are primarily conceived by those who operate them as opportunities for educating drivers about the dangers of speed (ACPO 2006, 2), it is apparent from drivers' feedback that one of the most important issues to be addressed by such courses is the provision of the opportunity for drivers to have 'voice' and to appear as fully contextualised human beings within the system. SACs allow drivers to discuss the individual circumstances of their offence in a situation presided over by a representative of the prosecuting authority, with some even inviting attendees to discuss their individual excuses for their offence. Although attending such courses secures a change in the penalty, this is dependent on successful completion of the course, not on the communication of mitigation or excuses that lead to the dismissal of the case. As such, the sharing of excuses cannot ultimately lead to the exoneration of the driver, but, as Tyler again notes, this is not a necessary feature of the opportunity to voice (1990, 127). Although 'pre-decision voice' is the most effective way of securing outcome acceptance, in the absence of this (as in cases enforced by automated technologies and strict liability principles) 'post-decision voice' is still valuable in securing satisfaction based upon fairness criteria (MacCoun 2005, 192). Voice, according to this finding, has beneficial qualities in and of itself and is not solely associated with the potential to 'get away with it', as these attendees suggest:[5]

5   It should be noted, however, that 'voicing' opportunities should be used to supplement otherwise procedurally just policies and not as a substitute for fair and just treatment elsewhere in the process. Where 'voice' is offered in isolation, it can be seen

It was nice to talk to the man from the Partnership. I don't feel that I should have been charged given the circumstances and I think he had some sympathy with my case. It cost £90 to find that out but at least I got to say it! (Female detected speeder, mid 50s, Speed Awareness Course)

The woman said that if I had complained at the time, they might've let me off it. I mean, it's obviously too late now but it was good to actually explain that the signs were just too close to the roundabout. (Male detected speeder, mid/late 30s, Speed Awareness Course)

SACs also present an opportunity for the driver to appear within the system as a person, and as a fully contextualised offender. As Tyler suggests, the opportunity simply to discuss the circumstances of the individual case, to appear as a respectable and polite individual and to feel that the authorities have been made aware of any concerns is valued by drivers. It is conceivable that, having vented their objections in a venue such as this, and perhaps having their justifying excuse challenged, drivers will be less inclined to vocalise their techniques of neutralisation in other arenas (the pub, the dinner table, the media, etc.).

The value of the opportunity to 'voice', however, remains a *side effect* of the creation of a situation in which drivers and Partnership representatives are brought together, rather than an intentional aspect. Future course design should therefore be careful not to remove this appreciated aspect accidentally by minimising the free input of attendees, even if these contributions do not appear directly relevant to the issue at hand.

A more procedurally just system therefore contains opportunities for communicating aspects of context which are seen to determine drivers' culpability for the risk they are accused of posing to others. By offering this kind of educative alternative to the standard fine and points, the important quality of human contact could potentially be rendered compatible with large-scale automated enforcement campaigns. Given that course attendance is unlikely to be a realistic possibility for all drivers, the retention of the 'comments' box on the Notice of Intended Prosecution is a cheap and viable alternative, which, this research suggests, is likely to be interpreted as offering at least a limited opportunity to exercise 'voice'.

A significant component of a more procedurally just system as conceived by some speeding drivers is, therefore, that it contains opportunities for communicating aspects of context that are seen to determine drivers' culpability for the risk they are accused of causing. Although the potential to resist being labelled as an instigator of risk is not a necessary outcome of such an alternative system for the drivers considered here, for some the advantages of communicating with a human being *are* viewed in terms of the potential for 'getting away with it'. Such drivers are considered next.

---

instead as a method of 'manipulation and exploitation' by the authorities to encourage recipients to accept otherwise unjust policies (MacCoun 2005, 193).

A Fair Cop

The importance of the changed nature of the interaction between accused and accuser has been a common theme of the discussion above. Instead of a qualitative, interpersonal interaction between a human representative of the problematising authority and the problematised individual, enforcement by techno-fix instead reduces that interaction to a greatly simplified measurement of the driver's worth. This new measurement considers only one aspect of the subject's behaviour at one isolated instant, removing it from the wider context in which that behaviour takes place. The inability for context to be recognised and acknowledged by automated detection and prosecution procedures is seen to be one of its major drawbacks in securing the consent and support of those to whom it applies. Given that this interaction is partially dehumanised, it is significant that suggestions for a more acceptable system of enforcement often centre around the lack of human characteristics on the part of the enforcing technology, characteristics that are seen to be vital components of a procedurally just system. Significantly, procedural justice analysis has, to date, assumed that judgements of the fairness of a system are made at points of inter*personal* interaction, between judges, police officers and the accused. As such, this final section considers the ways in which systems that can guarantee 'fair' and non-discriminatory enforcement are met with calls for discrimination to be reinjected into the system in order for it to be seen as 'fair' and legitimate. Such discrimination is, crucially, reconceptualised as the kind of *discretion* that results from the use of 'common sense' and 'intelligence' – in short, with *human* qualities.

The change in the experience of criminalisation brought about by a concern with risk and an enthusiasm for techno-fixes has, it seems, produced nostalgia for the days of the human traffic police officer. The perceived disappearance of the police officer from public view has been met with criticism in many circles (Loader and Mulcahy 2003, 28–9) and the roads policing context is seemingly no exception. Drivers who participated in this research frequently expressed a nostalgic preference for the days when speed limit enforcement was carried out not by technology but by humans in the form of traffic officers:

> You just wouldn't ever get a policeman [*sic*] who'd bother with you for doing 45 in a 40 in the middle of the night on an empty road! It's just common sense! They'd maybe tell you off and send you on your way, but they would never think of booking you and that's the way it should be – sensible like. (Male, late 50s, professional driver focus group)

This rather peculiar nostalgia was also in evidence in Blincoe et al.'s 2006 research, where respondents noted that:

The limiting factor of speed cameras is the absence of human judgement. Cameras cannot assess if a speed is justified and safe given the conditions. (Female driver, 54, cited in Blincoe et al. 2006, 375)

Speed cameras do not and cannot make the same allowances a policeman [*sic*] might. On the section of road I was photographed I do not believe I would have received a ticket from a policeman. (Male driver, 43, ibid., 375)

The 'common sense' alluded to in these examples and those below is clearly associated with the police officer's power to dismiss a driver with a warning, to filter them out of the system if the circumstances are right. Such circumstances relate to the contexts of time, place and individual character. As the following drivers explain, being stopped by a police officer offers the opportunity for a rational discussion about the issue, the outcome of which may potentially be the avoidance of a fine and penalty points, as the first driver suspects and the second confirms:

The human contact thing that people go on about I don't agree with because I think there would be an even greater humiliation to be talked to by another human being about an error you've made and for other people to see you doing it. And I think people who say they think they would rather have another human being to talk to, I think it's because they think if they had a person, they might be in some position to bargain away from the punishment they are going to get. (Male, late 20s, experienced driver focus group)

Terry: I was stopped by a policeman years ago in my MG. At least you are able to have, like, an interaction with them, plead mitigating circumstances.

Laura: Exchange points of view!

Terry: Yeah, and if he still wants to go ahead and book you, fair enough, but at least you have the opportunity to say, 'Listen, the circumstances were this and this and this', whereas with cameras there's no chance. (Convicted driver focus group)

The potential for an interaction with another human being is, crucially, understood as containing within it the potential for certain people to be filtered out of the system and for the effective decriminalisation of people who, for reason of respectability or context, *do not deserve to be there:*

Jane: I went on a Speed Awareness Course and there was this older person, and he'd never had any points on his licence, never been in any trouble. He was speeding by only about 5 mph because his son was critically ill in hospital. And the woman at the course said that if he'd phoned up, he would have been OK and they wouldn't have made him come on the course, it'd be cancelled, but he didn't know that. So if that'd been a police officer, they wouldn't have stopped him.

Martin: It would never have got that far, would it?

Jane: He actually wanted to go on the course, but it's just that lack of a *person*.

Martin: Discrimination. They [cameras] can't discriminate, no reasons behind it. (Convicted driver focus group)

In this final example, the need for discrimination in a procedurally just system is made explicit. Discrimination is, in this interpretation, a positive element that allows for people who deserve it to escape criminalisation. For these drivers, 'desert' can be determined only through an appreciation of context. The police officer acts as a gatekeeper to the criminal justice system which, it is assumed, will function as a common sense filter. He [*sic*] can listen to excuses as a similarly fallible human and acknowledge good intent and genuine mistakes. Fundamentally, however, the traffic officer is also able to identify the respectable, the responsible and the upstanding citizen, a capability likely to appeal to drivers who see that their law-abiding credentials entitle them to a little leniency when they come into contact with the law. The following driver suggests that his appearance and demeanour, supplemented by the type of car he was driving, were factors in his escaping prosecution – factors only recognisable to a human enforcer of the law:

I remember once being stopped by this young chap. He was looking at my car, which was one of those big new Rovers, and he was obviously interested. So I said to him, 'Look son, I've only just picked her up and I was just seeing what she could do.' And he asked me how fast I'd managed to go! He could see I wasn't going to cause any trouble, you know? That I was just a bloke having a bit of fun. (Male, early 60s, experienced driver focus group)

The legitimacy of a system is, according to this analysis, determined not on the basis of its ability to enforce without discrimination or bias but on the basis of the type of people it criminalises and consequently its inconsistent ability to demonstrate discrimination when appropriate. The system is judged not on the accuracy with which it detects offences but on the accuracy with which it seems to criminalise the 'right' people, as the following driver confirms:

It's the place where the boy racers are. I was down there with my mates for work and we said, 'This is where they should have the cameras, not on the A34', to catch these loons. And then that's where I was got! It was like 'Not ME', I'm the one, the 'concerned citizen' if you like, who wanted the cameras here, not the one they were supposed to catch! (Male detected speeder, mid 30s, Speed Awareness Course)

A contributor to Corbett and Grayson's research summed up the perception of excessive consistency and impartiality offered by speed cameras in the statement that: 'Cameras may not lie but they do not tell the whole story' (2010, 6). In short, the demand from some drivers is that all kinds of biases and discrimination be put back into the system, which is experienced as too impartial, neutral and consistent. The superiority of the traffic officer is seemingly that he [*sic*] can instinctively tell a respectable driver from a genuine 'troublemaker' and act appropriately, treating the two types differently and accordingly. The antecedent concepts of consistency, neutrality and impartiality provided by Tyler and colleagues are thus not linked in any simple way to the perception that a policy is procedurally just, and can, in some cases, actually function to reduce the perception of procedural justice. Human frailty and its potential for exploitation is, in the end, preferred to the type of guaranteed non-discrimination that the techno-fix can offer. A legitimate and fair system, it seems, punishes some ('bad', 'dangerous') people and contains the potential for other ('good', 'respectable', 'law-abiding) people to be filtered out, even when they have engaged in the same behaviour. The way in which speed cameras have been deployed is viewed as fulfilling neither requirement, given, first, its limited focus on one type of behaviour and, second, its persistent denial of the importance of situational and individual context.

As such, the notions of 'common sense', 'discretion' and 'respect' are considered vital to a 'just' experience. The fact that these are conditions of *human* enforcement and of inter*person*al encounters, and are therefore inapplicable to human–machine interactions, represents a challenge for future enforcement based around infrastructure of this nature. Such positively essential discrimination, voiced here as 'common sense', is attainable only through the use of human enforcement technologies. Whether or not this is a desirable goal is debatable, but the following examples explore adaptations to the basic speed camera system that may go some way to addressing concerns of this type.

Average Speed Cameras

Speed limit enforcement using Gatso-type cameras has often been criticised for capturing isolated moments at isolated locations, which itself is a major factor in some drivers feeling that their whole self has not been fairly represented or judged. The positioning of cameras has also been subject to intense criticism and is often the source of delegitimisation, as shown in Chapter 6.

Further, the propensity of drivers to slow down for speed cameras and then to speed up again has been noted by drivers, road safety groups (Transport2000 2002), motoring groups (see *Daily Telegraph* 2010d), academics (Corbett and Simon 1991; Keenan 2002) and even the government itself (DfT 2005a).

With this in mind, the potential for speeds to be controlled over greater distances is worthy of consideration. Some systems using specific types of cameras are able to calculate average speeds by measuring the time taken for a car to travel between two set points. If a vehicle is detected travelling at a speed that exceeds the limit plus whatever threshold is in use, a digital photograph of the offending vehicle is recorded. Penalty notices can then be generated by computer. As such, drivers are encouraged to keep to the speed limit for the duration of the measured journey.[6]

While this type of system has the potential to convince drivers that a more comprehensive evaluation of their driving has taken place, it too emphasises the importance of the limit as an indicator of safe behaviour. The driver is encouraged to concentrate on being compliant for the whole distance (by means of their speedometer) rather than on adapting their speed to the circumstances of the journey.

The additional cost of this kind of system was, in the past, viewed as an obstacle to its installation in many cases. The notion that speed enforcement should be self-financing (as apparently inherited from the hypothecation associated with Gatso-type enforcement) meant that many partnerships ruled out SPECS[7] systems during the period of the NSCP, fearing that the system would be too costly. Interestingly, the possibility that Partnerships would deter so many speeders that they would lose money is also phrased in terms of this being a 'risk':

> *HO*: SPECS isn't really taking off because it's still too expensive, and even if you hypothecate, you may not get your money back.
>
> *HW*: It works *too* well?
>
> *HO*: Exactly, the risk is quite high. (Interview with senior Home Office official 2004)

Consequently, the financial arrangements for speed camera revenue, so often a source of criticism in the wider debate, at one point seemingly also damaged the chances of technology being deployed in ways that had the potential to be construed as 'fairer' and thus more acceptable.

In more recent years, the use of average speed cameras, on motorways particularly, has become a more common sight, although no large-scale evaluation of the use of average speed cameras is yet in the public domain. Some evaluation of a limited number of camera pairs took place as part of the NSCP evaluation in 2005, with it being noted that 'time-over-distances cameras', as they are described here, are 'particularly effective at reducing excessive speed (more than 15 mph over the limit)' (Gains et al. 2005, 25). The DfT notes that:

> [e]vidence from the use of average speed cameras shows that they are effective in reducing speeds over longer stretches of road. A number of highway

---

6   The alternative is, of course, reducing their speed dramatically, or even stopping, at some points to make up for breaking the speed limit at others and therefore to generate an average speed below the speed limit.

7   Speed Enforcement Camera System.

authorities have submitted before and after evaluation data to the Department and this suggests reductions in the rate of KSI and reductions in the percentage of vehicles exceeding the speed limit have taken place at each of the sites. (DfT 2009d)

However, given that the data had not been validated independently, nor had it been adjusted to take account of the regression to the mean effect or national trends (ibid.), the DfT, in 2008, noted its intention to 'work to promote good evaluation of the latest generation of time-over-distance cameras and share the results with road safety stakeholders' (DfT 2008, 26). Seemingly, some of the lessons of the evaluation of the NSCP have been learned and their criticisms acknowledged.

It is difficult to gauge public reaction to this kind of technology specifically as anything with the potential to detect speeds is often simply described as a 'speed camera', without any further differentiation between types. The majority of this research took place before SPECS-type cameras were operational in significant numbers and they did not feature in participants' comments specifically. Many systems appear on motorways where roadworks are being carried out and, as such, may be seen by drivers as more legitimate as they relate specifically to the protection of those working on the roads. However, the first network of average speed cameras on a major non-motorway network was scheduled for installation in 2010[8] and, should this become a more common feature on the road network (particularly in the light of the withdrawal of government funding from the existing systems of cameras), further research could usefully explore any differences in driver responses to this alternative system.

Vehicle Activated Signs (VASs) and Speed Indication Devices (SIDs)

VASs and SIDs are devices located by the roadside that measure a vehicle's speed and display it on a screen for drivers to see. They are usually activated by a speeding vehicle and flash up a speed in excess of the posted speed limit. SIDs display either a smiling (☺) or sad (☹) face to the driver, depending on whether or not their speed is within or in excess of the limit. They are not connected to enforcement systems and therefore only have an educational or warning function. From 2004 to the end of the NSCP in 2007, these devices could also be funded by speed camera revenue (DfT 2004d, 45) and they can be installed in areas where the placement criteria for cameras are not met. As such, they can be placed in communities where concerns over speeding motorists have been raised but where cameras cannot be justified by KSI data.

Technologies of this type have been identified as a preferred option by some drivers, with one study of VASs citing 'overwhelming approval' of the method, with drivers of the opinion that they were installed with the aim of slowing

---

8   A network of 84 cameras covering a 7.5 mile stretch of the A13 in London (see BBC 2010c).

them down (Winett and Wheeler 2002, 17). Some participants in this research considered them to have an effective 'shock value':

> *Tom*: That thing Darren was saying about where the speed lights up. They were the best I've ever seen, they are really good.
>
> *Darren*: They really shock you.
>
> *Tom*: They shock you, so put that in your report, we think that's the one!
>
> *Darren*: It even flashes up what speed you're doing, and you think, 'This isn't just flashing for any old reason, this is actually telling me what speed I'm doing here', and if you slow down you can see it change. (Professional driver focus group)

Winett and Wheeler's research concluded that, in addition to having a substantial positive effect on accidents [*sic*], there was 'no evidence that in time, drivers become less responsive to the signs, even over three years' (2002, 17). However, in relation to the use of SIDs, Sadler's research found that, although they were effective at reducing speeds while present, there was some 'novelty effect' which meant that speeds decreased by smaller amounts in the second week of use than in the first week, and that after two weeks speeds began to increase despite the SID still being present (Sadler 2008, 3–4). Similarly, '[w]hen SIDs were removed, speeds returned to the speeds recorded before the SIDs were operational' (ibid., 3). Best practice recommendations included in Sadler's research therefore recommend that, for maximum effect, SIDs should be rotated around a number of sites (ibid., 4).

VASs and SIDs may be viewed as demonstrating a willingness on the part of the authorities to accept that speed limit breaches can result from momentary lapses. Drivers are treated as 'educatable' human beings, not 'mere fragmentary activators' (Lianos and Douglas 2000, 107). They provide drivers with the opportunity to modify their behaviour in order to return to a law-abiding status, and offer a second chance to learn from their mistakes. The technology treats the driver as a person who has lapsed and will be shamed into compliance, a potentially respectable person who will appreciate the opportunity to reform. As such, they may appeal to 'respectable', 'law-abiding' drivers who feel their offending is unintentional and thus undeserving of punishment.

However, it might be questioned whether the attraction of this method is actually due to an appreciation of its educational potential, or whether the appeal stems from the knowledge that the sign is not linked to enforcement technologies and effectively allows the driver to travel at whatever speed they choose without fear of prosecution. The support for their use evidenced by Winett and Wheeler's research, however, was found both among drivers who assumed prosecution *would* result from triggering the technology and among those who did not (2002,

16). As such, it should be concluded that such devices are viewed by many drivers as a more legitimate use of technology in the road safety context, being more respectful of drivers and more realistic about the practice of driving. They also have the potential to contribute to the ending of the notion that the detection and deterrence of speeding is inevitably linked to revenue raising, given that no fine income is generated by their activation.

Automatic Number Plate Recognition (ANPR)

ANPR cameras do not detect speeding motorists. However, their inclusion here is justified on the grounds that, in many respects, they are a similar technology to speed cameras and are directed towards the problematic behaviour of the same population – drivers. Despite these ostensible similarities, the response to the introduction of the technology towards the latter stages of the period of the NSCP has been markedly different to that of speed cameras. The reasons for this difference are explored briefly here.

Many forces now operate ANPR systems which, through the use of mobile camera units, can detect vehicles that are unregistered, untaxed and uninsured. The use of this technology has been noticeably less controversial than the similar technology in use for detecting speed limit infringements, and this section will suggest some reasons for this disparity, from which recommendations for the future handling of the speed camera policy emerge. At the height of the speed camera debate, one senior government officer noted that: 'There's been few press releases saying that ANPR is being rolled out now, but it came out and it was very quiet actually. We were quite surprised that *The Sun* didn't pick up on it and say, "You can't drive around anywhere now without somebody watching you"' (interview with senior Home Office official). It will be proposed here that this uncontroversial reception is a consequence of some of the themes identified as applying almost uniquely to speeding offences which have been identified in this research. The types of offence detected by the two technologies (ANPR and speed cameras) are seen to be of crucial importance to this explanation, with the types of offences being concentrated upon implicating certain 'types' of driver. While many speeders object to the emphasis on their behaviour, it is often within the context of the apparent neglect of 'real' motoring offenders such as the uninsured, the untaxed or dangerous driver. Drivers are seemingly aware of the fact of risk on the roads, but tend to think about this risk as external – as emanating *from other people* and impacting negatively upon themselves.[9] As such, they are more likely to see ANPR as a protective technology that limits the risks *to which they are exposed* on the roads. These risks include the chance of being hit by an uninsured driver or being in a crash caused by an unroadworthy (not MOT'd) vehicle, as

---

9   This is reinforced by the advertising of car safety features emphasising the crashworthiness of vehicles, while deprioritising and de-emphasising the impact of the car on pedestrians and other bodies outside of the car.

well as more generally being disadvantaged by other drivers' non-payment of road tax, for example. In the case of speeding, drivers are more likely to view fast drivers as benefiting them by not obstructing their progress (with *slow* drivers often receiving more criticism for 'holding up' or 'getting in the way' of faster drivers). Despite the increased risks associated with it, speeding is therefore perhaps the only offence on the road that is seen to *benefit* other drivers, rather than disadvantage them or expose them to risk.

Furthermore, the fact that a majority of drivers speed, but that only a minority of drivers fail to tax, MOT or insure their cars, means that ANPR technologies allow the majority of drivers to maintain law-abiding and respectable self-images and to continue (reassuringly) to view criminals as other people, reinforcing the essential difference between good and bad drivers, and good and bad people. As the Home Office official quoted earlier also noted:

> There's been very little press coverage about it. We think as more and more people are getting stopped they'll start asking more questions, and start asking the ABD for their view and all that, but it's seen as not just a motoring offences tool, but a real crime tool. It's working, it definitely works. The police themselves are amazed at the results because people generally without tax and insurance are more likely to be criminals – there has been research which has proven that. The potential is massive really. (Interview with senior Home Office official)

> We get a lot of letters from people mistaking ANPR cameras for speed cameras, and they've got no markings on them because the police don't want people to know they are out there. So people don't know what they are and they think they're speed cameras, because it's got a camera, but they needn't worry because it can't detect speed. (Interview with senior Home Office official)

The 'average' driver is therefore considered not to be at risk of detection by this technology, and therefore should not fear it. However, deliberately marketing ANPR cameras as 'not speed cameras', as seems to have been the strategy, could appear as a lack of joined-up thinking by those responsible for roads policing, although clearly the unpopularity of speed cameras in some circles has been the motivating factor for this. While lessons can and should be learned from the speed camera experience, throwing them to the metaphorical lions in order to avoid tainting future developments with their controversy is ultimately counterproductive.

A representative of the Department for Transport also made the point of associating these 'bad' drivers with other, topically relevant, 'real' criminals:

> ANPR can detect the presence of an illegal vehicle or a car that's foreign or under suspicion. And when they stop a vehicle that ANPR says is suspicious, it'll be not taxed, or not insured, but often it's also got someone in it that shouldn't be there, a bunch of drugs in the back, weapons, guns, it's all happened. (Interview with senior DfT official)

In this way, ANPR is being consciously presented as leading to the detection of 'real' crimes and as a way of 'denying criminals the use of the roads' (ACPO 2005). Figures for those arrested as a result of the use of ANPR, provided by a 2004 evaluation, and often cited in press releases and news reporting on the technology, included 2,263 arrests for theft and burglary, 1,107 for drugs offences and 1,386 for auto crime (Home Office, ACPO and PA Consulting 2004, 6). This continues to reinforce the established link that maintains that traffic offenders tend also to be deviant in other ways (Corbett 2003; Rose 2000). Road traffic offences are risk factors that indicate other criminal activity is likely, with ANPR publicised as offering a method of identifying these factors, leading to the discovery of more serious crimes. The public is, potentially, also more likely to accept this kind of expert research input when it maintains the police's control focus on the behaviour of an accepted group of 'real criminals'. Issues of the 'fairness' and 'justness' of the technology and of the judgements it makes therefore do not seem to surface when those it detects are perceived to be from the traditional criminal ranks.

In terms of recommendations to emerge from this comparison, it seems that the promotion of speed cameras could usefully draw on the successful introduction of ANPR. If the potential risks posed *to* drivers by other motorists speeding were emphasized, it is conceivable that speeding could be rendered more socially unacceptable. The message from the ANPR case is that drivers are more willing to accept surveillance and regulation of activities that impact negatively upon them, and which they associate with being engaged in by other people (Wells and Wills 2009). The current focus of speed educational campaigns on *pedestrian* injuries and deaths, while attempting to responsibilise the driver, neglects the fact that drivers, too, can be killed and injured by their own speeding and use of inappropriate speed and by the speeding and inappropriate choices of others. Future promotion of speed-related issues could therefore stress the potential effects on the driver, as well as on others, and emphasise to drivers that inappropriate behaviour by others puts them at risk.

Corbett also suggests other strategies that could contribute to the perception that 'law-abiding' drivers are being protected by, and not victimised by, roads policing methods. She suggests that, in addition to ANPR, the toughening-up of attitudes to and enforcement of licensing, registration and insurance legislation could contribute to the perception of 'legitimate members of the driving community' that they are not 'easy targets' while 'real criminals' escape the authorities' attention (Corbett 2008, 139). The suggestion that the recent ending of central funding for speed cameras (explored more fully below) was also, in part, designed to encourage local authorities to adopt a broader-brush approach to road safety could also contribute to an improvement in drivers' perceptions that a more rounded and context-appropriate approach to road safety was being adopted. It is conceivable that well-placed, well-justified speed cameras could then form an accepted part of a more varied portfolio of road safety interventions.

The End of the Road for Speed Cameras?

The complete demise of the speed camera was forecast by *The Independent* in 2010, with similar themes picked up across the media in response to the coalition government's decision to withdraw central funding from Partnerships. This research has suggested that the complete removal of the infrastructure supporting camera use is probably ill-advised and certainly unnecessary. However, the decision is likely to have significant consequences for the use of speed cameras and to bring about changes in the ways in which drivers experience centralised and local efforts at improving road safety.

Although it is perhaps too early to map the specific consequences of this policy shift, some indications are already becoming evident, with some councils publicising the fact that, without central support, they will be obliged to switch off their cameras.[10] Analysis of the response of the media, drivers and practitioners to this one example suggests some patterns for the future.

First, it should be noted that the ending of central funding has not meant the ending of the accusations of 'revenue raising' that have dogged the use of speed cameras since hypothecation was first introduced in 2000. Instead, in a move that the coalition is unlikely to have predicted, suggestions have been made that the government is now making a profit from speed cameras for the first time as it is no longer making a financial outlay, but continues to receive the fine income (*Daily Telegraph* 2010c). This directly challenges an expressed aim of the decision which was made in order to 'end the war on the motorist' and ensure that they were no longer 'cash cows' for government (widely reported, but see, for example, *Independent* 2010a; *Daily Mail* 2010b).

Naturally, many comments on the withdrawal have raised concerns about road safety, with the ACPO roads policing spokesman expressing fears that road crashes will increase. These views are presented alongside support from road safety charities (BBC News Online 2010c) and are logically a central criticism of the decision, given the well-publicised improvements in road safety attributed to cameras. However, the narratives surrounding the announcement to withdraw funding for cameras from local authorities make no mention of the 'effectiveness' argument, focusing instead on the issue of acceptability and noting that the decision signals the coalition government's commitment to ending the 'war on the motorist'. This suggests that it is public opinion that has been the key driver behind the change in policy, although suggestions that the new arrangements actually increase the revenue received by the government have, as noted above, continued to surface.

---

10   Oxfordshire County Council was the first to do this, while Wiltshire County Council and Buckinghamshire County Council both dismantled some cameras (*Daily Telegraph* 2010d). On 1 April 2011 Oxfordshire County Council switched its cameras back on, citing worsening crash data as their reason for doing so (BBC News Online 2011).

A further justification offered by the coalition government has been that the move is designed to encourage local authorities to consider a wider range of interventions, including education, signage and engineering, rather than considering the speed camera to be the default option when a road safety issue presents itself (BBC News Online 2010d). While this would be likely to appeal to motorists who claim that speed cameras are a blunt instrument applied to a complex problem, such has been the rhetorical purchase of the 'revenue raising' implication that this potentially more acceptable stance is effectively overlooked.

Media reporting veered between expressing concern at the potential for the removal of cameras to lead to increased crashes and the reporting of data that suggested that the removal of cameras has led to a decrease in crashes – sometimes within the same month in the same newspaper (see *Telegraph* 2010e, 2010f). The results of an Oxfordshire County Council speed survey, conducted at only two, then decommissioned, camera sites over the five days that had elapsed since the cameras were switched off, was also reported as though offering conclusive proof of the effectiveness of cameras at reducing speeds and the apparently inevitable increase in deaths and injuries that will result from their removal. In responding to the findings, the road safety charity Brake again focused attention on the core risk underpinning the debate in noting that 'this is people's lives we are talking about' (ibid.). However, other reports in other locations, released within days, suggested that cameras have been responsible for 28,000 crashes in the decade since their installation. The ABD, summoned to offer a response to the research, notes that distraction of drivers by the threat of prosecution is inevitable, and demands that '[w]e need to get police back on the roads. Cameras cannot see if a driver is driving dangerously or on drugs but a police officer can detect that' (*Daily Express* 2010b). The debate continues, with the same expert players vying for definitive status using the same arguments, even as the suggestion is made that it is 'the end of the road for speed cameras' (BBC News Online 2010e).

In acknowledging that speed cameras may be unpopular among some populations, and acting on this observation, the present government has, unlike its predecessor, recognised the existence of a debate about the use of speed cameras and of disquiet among those who fall within their gaze. Their response has been to distance the government from the policy by 'localising' the issue and ending central funding for speed cameras, 'empowering' local authorities to choose whether or not they wish to continue to fund them or to substitute other road safety measures (DfT 2011).

The findings of this research suggest that a more productive move would have been to acknowledge the concerns that constitute oppositional narratives but to respond by seeking to address them in ways that do not undermine the technology nor sidestep the issue of its effectiveness. This chapter has attempted to explore ways in which an acknowledgement of debate around the subject of speed cameras need not necessarily lead to their demise, but can instead lead to constructive development of the policy in directions that are likely to increase both its effectiveness and acceptability. Although it has focused on the oppositional voices in respect of speed cameras, this research also met considerable support for the method, and a more constructive future for automated speed limit enforcement is likely to emerge from a stance that does not seek to deny the existence of either position.

A New 'Twin-Track' Approach to Traffic Offending?

As this book was nearing completion, the coalition government launched its *Strategic Framework for Road Safety* document (DfT 2011). Central to the new strategy was what a DfT spokesman described as a 'twin approach' whereby, in Transport Secretary Philip Hammond's words, the intention was 'to crack down on the most reckless and dangerous drivers' and 'support those who are basically law-abiding but perhaps have an occasional lapse' (BBC News Online 2011b). This was to be achieved through policies 'focused upon making it easy for road users to do the right thing, while taking a tough approach to those who deliberately decide to undertake antisocial and dangerous driving behaviour' (DfT 2011, 18). A range of disposals would then divert those who made 'low-level' mistakes into education courses, with tougher enforcement reserved for the 'minority' who 'commit serious, deliberate and repeated offences' (ibid.). While such rhetoric may well, on the surface, appeal to drivers who argue that there are different 'types' of road user and different 'types' of traffic offence, it is not clear how these distinctions could be made in practice in any way that accurately reflects the reality of risk on the roads. Policies and practices designed to 'encourage people to make the right choices' (DfT 2011, 46) are not compatible with laws designed to reduce risk and which are established on strict liability principles given that, necessarily, even unintentional behaviours are deemed punishable. Similarly, while the construct of the 'accidental/low-level/otherwise law-abiding' offender versus the 'more serious/deliberate/repeat' offender may appeal to many road users, it must ultimately be unsustainable, given that serious offences need not be deliberate and even minor unintentional lapses can be fatal. However, this does seem to mirror what many drivers themselves seem to understand as representing the reality of road-user behaviour which, it seems, is divided into either 'real' criminal or law-abiding drivers, and 'dangerous' as opposed to unintentional offenders. There also seems to be evidence of an intention to bring in a more contextualised response to illegal driving of the sort detected by ANPR, with the suggestion that '[w]e are concerned that wilful and repeated acts of non-compliance by this group will reduce the faith that the responsible majority have in enforcement and it may erode their own will to comply' (ibid., 63). Again, such an approach is, on the surface, appealing to many drivers who objected that they were 'easy targets' (Corbett 2008, 139) because of their law-abiding approach. Again, however, the compatibility of these aims with automated detection systems is not entirely clear given that establishing what is a wilful act of law breaking is not something that ANPR, speed cameras or any other out-of-court system based on strict liability or fixed penalities is set up to recognise or acknowledge in any way.,

Wider Implications and Future Research Directions

This work has considerable scope for future application in a society in which control, regulation and punishment are increasingly justified on the grounds of risk. There are, I believe, a number of possible directions for future research, some of which retain a focus on the traffic context and some of which suggest a broader focus. The common themes of such research would, however, mirror the approach of this work by centring around the notions of truth, identity and justice in such a society and the ways in which each is shaped and altered by a concern with risk.

Future research could therefore usefully consider examples of so-defined risky behaviour that demonstrate the potential to be tackled in ways that are in evidence in the case of speed limit enforcement. The deployment of strict liability principles on the basis of risk assessments, enforced by techno-fixes within fixed penalty structures, is likely to be an increasingly familiar scenario for the future. Strict liability laws are the logical response of a society concerned to minimise risk but faced with regulating behaviour that caused risk whether or not it was intended. Techno-fixes are a feasible and cost-effective method of detecting when such laws are breached and fixed penalties are a viable method of punishing such offences. The potential side effects of such developments, as demonstrated here in the case of speed limit enforcement, are therefore particularly apt subjects for research with a wide and potentially growing application and relevance for society.

A number of issues will therefore be considered briefly here which demonstrate some or all of the characteristics emerging as significant in relation to the speed camera debate. Such an agenda produces a varied list of activities both within and beyond the traffic context, including failure to tax, license or insure a car, the use of mobile telephones while driving and (following proposals in the 2011 *Strategic Framework for Road Safety*) careless driving (DfT 2011, 10), as well as the effects and control of smoking in public places. It also produces a variety of offences broadly consumed within the title of public environmental issues or issues of 'local environmental quality' in the language of the Department for Environment, Food and Rural Affairs (DEFRA). These include littering, dog fouling, noise and light pollution, abandoned vehicles, illegal parking and graffiti. Although apparently disparate, the issues contained within this list share certain characteristics. Not only are they increasingly conceptualised as 'risk' issues but they demonstrate the potential to be conceived of in strict liability terms, are amenable to the use of a techno-fix in some form, and are punished outside the traditional theatre of the courtroom by means of the fixed penalty. Perhaps most importantly, however, they are all issues that are of particular interest to the public who are constructed as both the 'at risk' victim and the problematic 'risky' individual in such cases. Risk issues with some sort of public ownership are, this research has suggested, likely to be those particularly prone to debate, with the public having both a vested interest in the outcome and a degree of experientially derived expertise about the issue.

Following the example of the debate that forms the subject of this research, it might be expected that disagreements about causal interpretations and disputes about the legitimacy of enforcement could arise in the case of many risk issues. The debates that have taken place about the restriction of smoking (an offence for which a fixed penalty has also been deemed appropriate punishment) demonstrate that risk debates, in the context of demonopolised expertise, are not confined to the road traffic context. Debates stimulated by the initial proposal to introduce a ban on smoking in enclosed public spaces across the UK did, in some senses, mirror those in evidence in relation to the use of speed cameras. Similarities include the existence of pressure groups which campaigned against the existence of causal linkages between passive smoking and physical harm, including the smokers' lobby group Forest. Their objections are centred around the claim that the evidence used to justify the ban on the grounds of risks to non-smokers' health are 'based on nothing more substantial than estimates, guesswork, subjective recollections and even gossip' and that 'the dozens of studies conducted around the world over the past 25 years fail spectacularly to yield any reliably stable, uniform or statistically significant link between lifetime exposure to environmental tobacco smoke and lung cancer in non-smokers' (President of Forest, quoted in BBC News Online 2005g). Those in support of a ban, however, acknowledge it as being 'the most important advance in public health since [it was] identified that smoking causes cancer 50 years ago' (Professor Alex Markham, Cancer Research UK, quoted in *Daily Mail* 2006b). In other press reporting of the House of Commons vote on the introduction of a ban, such conflicting expert 'truths' were often juxtaposed, presenting the issue of risk and smoking, just like that of risk and speeding, as a contested issue (see, for example, *Daily Telegraph* 2006). The similarities between these arguments and those employed by those against the use of speed cameras to enforce speed limits are obvious and again create a scenario in which the justifications for the ban appear subject to dispute in the absence of an apparent expert consensus. In a society in which expertise has become demonopolised, those potentially negatively affected by regulatory interventions can draw on expert sources to evidence opposition to the ban, while those in favour can cite experts who support the causal link between passive smoking and damage to health. The regulation of individual behaviour with a view to the minimisation of risk is, once again, the outcome that is being fought over in this scenario.

An area in which regulation has often followed strict liability principles and in which the language of risk is commonplace is that of environmental harms. Environmental damage is increasingly coming to be seen as an individual problem as well as an issue for big businesses and states. Issues such as smoking, littering, noise nuisance and dog fouling are increasingly coming within the scope of regulation justified on the grounds of their environmental impact and potential health implications. In all of the above cases, the appropriateness of the use of strict liability principles for problematising individual actions has been discussed. Irresponsible attitudes to the disposal of plastic bags that result in environmental pollution have, for example, been punished as strict liability offences (Scottish Parliament 2005), while irresponsible dog owners could also have encountered such principles if they failed to dispose of their pet's environmentally hazardous product appropriately (Hansard HL 1995). Creators of environmental harms through the generation of noise nuisance are equally likely to encounter strict liability principles (McManus 2000), while being in control of a burning cigarette in an area where smoking is prohibited potentially embroils the guilty party in such legal procedures (Maxwell 2004). The potential extension of such principles into these areas further extends the reach of strict liability laws into the general public sphere where concerns about the compatibility of such principles with justice have, as has been shown, been raised.

Fixed penalties have been viewed as appropriate for offences of truancy, disorder and the selling of age-restricted products, among others, where they are considered to send 'a strong signal by delivering an immediate and visible penalty' (ibid., 32). Furthermore, legislation passed in 2005 encouraged the wide use of fixed penalties for controlling such 'antisocial' behaviour as graffiti, dog fouling, 'nuisance parking' and litter (DEFRA 2005). Little research has considered the experience of receiving such a penalty (though see Amadi 2008), focusing instead on the perceptions of those issuing the penalties. This research suggests that the expansion of the use of the method without a more detailed understanding of the recipients' beliefs about and understandings of the process could lead to complications. Public opinion may be divided both in respect of the targeting of 'real' (antisocial behaviour, graffiti) versus 'minor' (dog fouling, littering, parking) offending, as well as between those suffering the effects of and those being responsible for these latter offences. Recent controversies following the decriminalisation and privatisation of parking enforcement have also shown similarities with the speed camera debate, with research suggesting, for example, that recipients of parking penalties often choose to 'get personal' and take their case to a tribunal where they could have their case heard (Raine, Dunstan, and Mackie 2003).

The potential for such risky behaviours to be punished via the use of fixed penalties means that an increasing range of problems come to be punished in ways that circumvent traditional court-based justice systems. This has caused some magistrates to express concerns at the increasing use of the fixed penalty method and its inability to consider personal circumstances (Magistrates' Association 2010, 5). An increasing number of risky individuals are potentially going to be processed and punished in situations remote from the courtroom and from representatives of the criminalising authority. The consequences of such a modification in the experience of receiving punishment (with or without experiencing 'justice') have been raised here as subjects of potential concern. The consideration of procedural justice antecedents in relation to these newly configured relationships (and in the absence of the interpersonal interactions in which such encounters have always been assumed to take place) is thus an innovative approach with considerable topical relevance.

Continuing the themes that this work has set out to illuminate, future research could also usefully include a consideration of the reception of a variety of techno-fixes. Throughout this research, comparisons with CCTV and ANPR systems were made but fell beyond the scope of this study. In the former example, the popularity of this intervention could be explored in terms of its operation, and therefore *mediation*, by *human* controllers – the fallible, biased and discriminating missing ingredient whose absence was bemoaned in speed camera enforcement. In the second example, the use of ANPR systems to detect illegal road-user behaviour (such as untaxed, unlicensed and uninsured driving) provides a valuable comparable context, given that its introduction has passed by without significant comment and certainly without the level of controversy that surrounds speed cameras. This is despite the ostensibly similar road safety context and despite the fact that such offences are also strict liability offences, punished by fixed penalty, and detected by a techno-fixing camera. Such technologies have been explicitly promoted as focusing on legitimate criminal targets and leading to the arrest of 'real criminals' including burglars, drug dealers and drink drivers (BBC News Online 2005h), persistent offenders (BBC News Online 2004c) and even terrorists (BBC News Online 2002b; Campbell and Evans 2006). The respectable driver is thus reassured that s/he is not the target of this enforcement. As such, this parallel development provides an opportunity for exploring the significance of the 'respectability' of the target population in generating approval of, or opposition to, a new intervention motivated by 'risk'.

The traffic context therefore appears to provide the most potential for the potent mix of the techno-fix, strict liability legal principles and the use of the fixed penalty, although all methods have experienced a growth in usage and provide potentially revealing focuses for future research. A preoccupation with risky behaviours means that strict liability regulatory offences based on *mala prohibita* presumptions are likely to increase in number, with techno-fixes and fixed penalties also likely to appear increasingly attractive as enforcement options. The replacement of policing as a human function with automated detection based on risk premises is, this research suggests, resulting in a reconceptualisation of the relationships between police and policed, also based on risk premises. 'Empirical studies of technology in use' (Haggerty 2004a, 493) such as this research are therefore crucial to understanding how risk-based interventions and the deployment of techno-fixes will impact on the lives of those at whom they are aimed and the roles of those who would seek to employ them and, perhaps most importantly, bring about changes to the relationship between the two. It is only through stepping outside of the ongoing debate about the use of speed cameras that the significance of such concerns for present and future crime control and governance, within a world in which both regulators and regulated increasingly organise their social world in terms of discourses of risk, can be realised.

Chapter 8
Conclusion

Beck describes 'risk determinations' as 'an unrecognized, still underdeveloped symbiosis of the natural and the human sciences, of everyday and expert rationality, of interest and fact' (1992, 28). This work has recognised such a symbiosis in the case of the debate around the use of speed cameras to enforce speed limits. This has been explored through the use of a variety of methodological approaches that have accessed both expert and everyday discourses, considered the combination of the natural and human sciences, and explored a context in which interest and fact are rendered virtually indistinguishable. This research has, furthermore, demonstrated the way in which a concern with risk both justifies the initial intervention in relation to speed limit infringements and determines the nature and form of the response to it. The foregoing analysis has therefore demonstrated how this debate has been influenced at several levels by a concern with risk and by its location within a society increasingly concerned with talking and thinking in 'risk discourses' (Hunt 2003, 183). 'Empirical studies of technology in use' (Haggerty 2004a, 493) such as this research are, it is proposed, crucial to our understanding of the ways in which an increasing concern with risk impacts upon both the experiences of governing and being governed.

This thesis has taken a deliberately non-partisan stance in relation to the most frequently asked question in relation to speed camera enforcement: 'Do they work or not?' It has done so in order to explore the debate that continues around this question, rather than contribute to it. The failure of persistent attempts to resolve such ostensibly simple questions is explained in two ways. First, such a failure is taken as an indicator of broader and deeper issues that the subject of speed enforcement raises for those who perpetuate the debate while at the same time trying to resolve it. The real questions underpinning the debate are, it is argued, those of truth, identity and justice, not those of technical effectiveness at all. Second, the existence and intransigence of this debate is also, it is proposed, explained by the risk context in which it operates. Such a context provides the means, motive and opportunity for a debate of this nature to take place.

Truth, Identity and Justice

This research has demonstrated that 'truth' is both increasingly *necessary* in a debate such as this, and increasingly *illusive*. The increasing conceptualisation of behaviours in terms of their 'risky' nature does not mean that it is no longer necessary to prove that such behaviours are also 'wrong'. Indeed, the demonstration of a connection between a behaviour and an identified harm is essential for the justification of the prohibition of that identified behaviour. It is just such an expert truth that is required to legitimate interventions and restrictions made in relation to *mala prohibita* offences. It is just such offences, furthermore, that are likely to be increasingly created by authorities keen to problematise behaviours that it views as causing risk but which are not pre-existingly 'wrong' or 'immoral' acts.

'An' expert truth is, however, harder to establish and constantly contradicted given that expertise has been demonopolised and dispersed across a variety of competing suppliers. As a result of this demonopolisation of their product, experts have to compete to get their particular interpretations accepted by the public audience who will be subject to them and upon whose compliance they are reliant. Experts previously legitimated on the grounds of their status are now challenged by interlopers questioning both their legitimacy and their honesty and must promote their supposedly objective scientific product in a marketplace full of contradictory scientific truths. The existence of competing 'truths' therefore creates an apparent dissensus around a risk issue such as speed.

Various methods of promotion are therefore used in an attempt to set one 'truth' apart from another. First, access to the media is considered crucial given that the audience for such marketing strategies is the general public, whose cooperation and consent is required. Given their vulnerability to claims of vested interest, experts are then seen to employ the services of 'independent' experts in generating the evidence base in support of their particular causal interpretation. Owing to the central importance of such scientifically derived proof, other strategies for promoting one's own evidence relate to the destruction of an opposing evidence base. This is achieved by attacking the credentials of an opponent in various ways, ways that suggest that they are deficient in their use of scientific methods and thus that their 'evidence' is not, in fact, evidence of anything at all. In this way, the existence of apparently contradictory 'truths' can be explained. Further strategies then promote the impression of an opponent's truth as not only marginalised and unrepresentative, but increasingly so. In this way, the existence of the debate in the past can be explained, but the impression that a consensus is building in favour of a particular causal interpretation can also be generated. As a parallel to this strategy, experts attempt to homogenise their audience by implying that a particular interpretation is, in fact, supported by 'common sense'. Such a reliance on common sense attempts to by-pass the apparent contradictory findings of scientific research by instead appealing to the audience on a more intuitive level.

These approaches are, however, rendered problematic by the risk context in which they are being deployed, and betray a misunderstanding of the nature of the public audience for the debate. First, each marketing strategy is, it has been shown, used by both 'sides' with equal confidence. As a result, no single truth is still able to emerge given that all truths are promoted with equal conviction. Second, attempts to invoke *a* commonsense consensus assume that this is something that can be constructed or formed *on behalf of* the lay public by simply claiming that one interpretation 'fits' while another does not. This second approach fails to recognise the increasing democratisation of expertise which, in the case of this risk issue, now includes the 'lay' public. Unlike many 'big' risk issues, driving is seen to be an 'everyday' risk issue (Hunt 2003, 167) and one about which over 30 million people in the UK alone can claim to have some kind of experience and knowledge. This allows those traditionally conceived as lay members of the audience to view themselves as experts with regard to road use and the causal linkages that operate in relation to speed and the likelihood of crashes. In this sense, the public audience for this debate is seen to be increasingly emancipated from dependence on such experts and their product. Instead, they are able, increasingly, to generate their *own* 'truths' as a result of their daily experience of, and experiments in, risk. This expertisation is alluded to by drivers who, in this research, often chose to describe themselves in ways that demonstrated their knowledge of driving and road-related issues. Such chosen identities frequently emphasised 'professional' driver status, or referred to a large number of years of experience behind the wheel, while particular viewpoints were supported with examples of apparent cause and effect drawn from their own experience of driving.

Regulation and control on the grounds of risk is also seen to pose significant challenges to the maintenance and defence of a coherent and reliable identity, just as it does to the maintenance and defence of a coherent and reliable truth. A discourse of 'identity' has therefore emerged as a second significant theme of this research. Rather than traditional categorisations based around gender, age, race or class, for example, this research has identified that other 'do-it-yourself' identities (Mythen 2005, 132) such as that of 'respectability' are used by some drivers. Such an identity is seen to situate the individual within the moral majority, is characterised by a law-abiding stance and involves making an active contribution to society. Such identities are, however, rendered irrelevant when problematisation is pursued on the grounds of risk. A risk-reduction motive permits the use of strict liability legal principles in relation to *mala prohibita* offences committed by the vast majority of the population and, as such, law-abiding, moral members of the majority can be criminalised and cast as immoral, deviant offenders. Both past behaviour and intended future behaviour are rendered irrelevant. Instead, the 'risky' or 'not risky' status of present actions at risk checkpoints such as the speed camera become definitive and justify the future control and punishment of the individual. Identities already reflexively constructed and in constant need of reinforcement are therefore undermined by risk-based interventions that care only for the risky dividual characteristic of their target population.

As a response to the challenges posed by such enforcement methods, a number of strategies have been used by some drivers to resist the negative implications and consequences that speed limit enforcement of this nature brings about. Such responses can be viewed as attempts at both responsibilisation and deresponsibilisation in relation to risk and rely on the accused driver reconceptualising themselves as the victim of risk, rather than the instigator of it. 'At risk' is the respectable identity and all that it symbolises and represents, and protection against this risk can be secured in a number of ways. First, the tactic of highlighting other harmful behaviours that are seen to be more worthy of enforcement attention achieves a degree of deresponsibilisation for the accused driver. Rather than accept an identity based around being an instigator of risk, and all the negative consequences that brings about, speeding drivers are able to recast themselves in the more reassuring and less identity-threatening role of potential victims of 'real' crimes such as burglary, murder or robbery. This is achieved by attempting to shift the focus of attention on to such other 'real' criminals. Such 'real' criminals are, crucially, viewed as being deviant individuals who commit *mala in se* crimes, for which intent is required, and which the 'respectable' would never commit.

Responsibility for road-based risk is also neutralised by blaming road crashes on other 'bad' drivers (drink drivers, dangerous drivers, even slow drivers). Such legitimate enforcement targets are, it is felt, then free to victimise 'good' drivers because the police have withdrawn from the streets and entrusted roads policing to the speed camera. Such methods achieve deresponsibilisation by attempting to reconstruct the driver as the victim of a greater road-based risk than that which they are accused of causing.

Second, and more directly, speeding drivers are also able to protect themselves against threats to their respectable identity by physically intervening to prevent their own detection and prosecution on the grounds of risk. Such methods involve drivers arming themselves with equipment, knowledge and tactics that prevent their 'victimisation' by speed cameras. By deploying such methods drivers can then avoid the negative labels and consequences that such problematisation would cause, and are instead able to continue to speed without risking their respectable identity or driving licence.

Concerns with risk mean that the approaches to control and regulation explored in this research are only likely to increase in popularity. The threats to identity that emerged as significant in this research are likely to be increasingly familiar experiences for the public when controlled on the grounds of risk. The methods by which implicated instigators of risk can effectively neutralise their responsibility for it are also therefore likely to present as challenges for those charged with effectively problematising risky behaviour.

'Justice' is, furthermore, seen to be under significant threat as a result of the specific methods used to problematise risky behaviours. The discourses of justice and fairness have, as a result, emerged as significant in the speed limit enforcement context. The use of risk as a motivating concern has allowed the use of enforcement systems, legal frameworks and detection methodologies that are concerned with only certain aspects of the offence and the offender; namely that an incidence of non-compliance with speed limits is detected and recorded and that an accountable individual is traced. The use of 'acceptable levels' frameworks to underpin enforcement, the deployment of inanimate objects to detect breaches of these levels, the use of strict liability legal principles and the ultimate punishment through fixed penalty systems demonstrate that the system is concerned only with this detection and accountability. The use of the techno-fix at the centre of such approaches, while providing what is potentially the 'fairest' of all enforcement (Lianos and Douglas 2000, 108), is seen to be capable only of providing a fair and just experience on a very basic level. Although such approaches to the control and regulation of risk provide unprecedented opportunities for 'consistency', 'impartiality' and 'neutrality' (Tyler and Lind 1988, 131; Tyler 1990, 7 and 117) given that they are incapable of discriminating on the grounds of age, race or gender, they are seen to neglect to give proper recognition to new identity categories that a system *should* acknowledge in order to be viewed as fair. Although 'law-abiding' and 'respectable' identities are frequently deployed in response to problematisation by speed camera, they are found to be of no interest to such systems. Similarly, identities that construct the individual as somehow above average at driving by virtue of being a 'professional', having vast amounts of 'experience' or otherwise just being a 'good driver' are also dismissed by technologies that utterly decontextualise offences and offenders. A neglect of such considerations is viewed in terms of an absence of 'respect' (Tyler 1990, 7) for relevant differences that make problematisation on the grounds of risk unjustified. Instead, the speed camera, and its associated approaches to legal problematisation and the dispensing of punishments, considers that all have posed the same amount of risk by engaging in the same behaviours. The guarantees of non-discriminatory enforcement that such approaches can boast are, therefore, in themselves contributors to the allegation of unfairness and injustice with which they are met. This apparent contradiction has been explained via the use of procedural justice approaches which demonstrate that the type of discrimination eradicated by such methods is actually necessary *if* those punished in this way are to accept their punishment as legitimate.

The injustice and unfairness of such an approach is exacerbated by the absence of an opportunity for the driver to appear as a fully contextualised human being at any point from detection to disposal. Many drivers were dissatisfied with the denial of opportunities to 'voice' throughout the speed limit enforcement process, and were critical of the camera's inability to be reasoned with, to show common sense or to exercise discretion. Beyond this initial 'detection' stage, the use of strict liability laws was seen to function to further dismiss relevant excuses and mitigating circumstances. Finally, the use of fixed penalty systems also denied accused drivers both their 'day in court' and their chance to speak to a human representative of the authority who had identified them as problematic. In this respect, this research is able to suggest that it is within the *interpersonal* aspects of a system that procedural justice is generated. While the existence of such human contact has, to date, been taken for granted in 'unfavourable outcome' (Tyler 1990, 107) encounters between the public and authority, this can no longer be assumed. Instead, a concern with risk has produced systems that can successfully detect and punish risky behaviours without the need for risky individuals to interact with human representatives of the authority problematising them. This research therefore suggests that although human contact can (in a strict, practical sense) be designed out of systems that provide efficient and effective risk detection, it cannot necessarily be designed out of systems that provide a just, fair and legitimising experience.

However, drawing again on procedural justice approaches, the absence of a perception of justice in such encounters with automated detection and processing systems is, it is proposed, likely to result in long-term reductions in the perceived legitimacy of those that deploy them. Reductions in levels of compliance and cooperation are, as a result, unlikely to be compatible with a desire to deter risky behaviour and thus bring about reductions in levels of harm. If a just experience is one involving human contact, and just experiences are necessary to ensure compliance with the system, the success of techno-fixes and their associated processing mechanisms at achieving measurable reductions in risk may, in fact, be limited.

Means, Motive and Opportunity

Through its location within risk society, speed camera operation policy is seen to provide the means, motive and opportunity for the controversy that surrounds it. The *means* by which official versions of the causal chains at work in relation to speeding can be rejected are provided by the contexts of demonopolised and democratised expertise. Being able to dispute the need for regulation at a most basic of levels is, the research proposes, a necessary first step for a debate of this nature, and the context of demonopolised expertise provides the expert dissensus that allows this to happen. Causal interpretations that both support the use of speed cameras and construct their use as illegitimate are available and are, furthermore, found to be scientifically 'proven' and validated by 'experts'. Individuals are, as a result, able to believe that speed limit enforcement is pursued where it is inappropriate and unwarranted and to find expert support for such a stance. Such a rejection of the legitimacy of enforcement would not, it is proposed, be as plausible if drivers were faced by an expert consensus in favour of the significant role of speed in crash causation.

The increasing expertisation of the general public also means that drivers are able to apply their own experiences when selecting a causal interpretation to subscribe to, endorsing interpretations that appear to support their own experiences and rejecting those that do not. For the individual driver who frequently speeds and seldom, if ever, experiences harm as a result, the reality of lived experience may appear to belie the causal statements of some experts and support the interpretations promoted by others. This expertisation of the public therefore equips the audience for the debate with further means with which to stage challenges to official interpretations.

The *motive* for engaging in this debate is, however, also provided by the risk context in which the debate takes place. The various expert interpretations of the versions of the truth in relation to the role of speed in crash causation do not, it is proposed, compete on a level playing field. Drivers have a vested interest in accepting some truths and rejecting others. Causal interpretations that justify unpleasant consequences for individuals in terms of punishment and restriction are understandably less appealing than those that allow individuals to continue behaving as they wish without the threat of punishment. The consequences of being defined *as a risk* are, as has been shown, magnified by the risk context that makes the adoption of causal interpretations disputing this definition more attractive. It is the potential consequences of such threats that therefore provide the motive for individuals to reject unpalatable causal interpretations that justify harmful restrictions and interventions, and to adopt ones that insulate them from such unpleasant consequences.

Finally, the use of the speed camera as the method of detection and problematisation is viewed as critical to the production of this debate, as it provides the *opportunity* – the specific opening or favourable set of circumstances – in which the wider objections to the appropriateness of enforcement can galvanise. While the existence of contradictory scientific 'truths' in risk society allows individuals to view enforcement as targeting the wrong *places* (where 'risk' was not 'really' caused), the need to maintain a coherent identity in the face of challenges in risk society allows individuals to view enforcement as targeting the wrong *people* (who, for various reasons, did not 'really' cause 'risk' either). The debate is therefore underpinned by concerns about the inappropriateness of enforcing speed limits in 'safe' places and against 'safe' drivers, with either the circumstances of the offence or of the offender considered not risky enough to warrant the negative consequences that resulted. Such objections are captured in the often-heard claims that enforcement is targeted at the 'safe driver', travelling at 'only a few miles over the limit' in 'the middle of the night', while the 'dangerous driver', travelling at 'twice the speed limit outside a primary school' evades capture.

However, it is the use of the speed camera itself as the technology of control and detection that provides the opportunity for such complaints to be deployed repeatedly and enthusiastically. Such objections, it is proposed, would not be heard as often if a different method of enforcement was used. This is demonstrated by the repeated calls for a return to humanised roads policing (with all its biases and potential for discrimination) that have been made frequently within this, and other similar qualitative, research. Some drivers see this type of enforcement as offering opportunities for contextualised enforcement that would problematise only genuinely and demonstrably dangerous or reckless behaviour. The frequently-heard calls for a return to the days when human traffic officers detected speed limit infringements can therefore be understood as a call for a return to the kind of 'common sense' and discretionary policing that resulted in only 'fair' prosecutions. Such prosecutions, it is assumed, would result from the traffic officer targeting only legitimate locations where speed demonstrably caused a risk (such as the 'primary school' example used above and frequently deployed in the debate) and/or targeting only 'dangerous' or 'bad' drivers who were clearly putting others at risk. The sorts of offences that cause controversy (the marginal offences on the 'empty roads in the middle of the night' used in the example given earlier) would, it is assumed, seldom be detected or punished by such a method. The vast majority of offences detected in this way would, it seems, be considered legitimate and the policy would offer less scope for criticism. It would, for example, be harder to construe oneself as the victim of speed limit enforcement if punishment resulted from a method of detection that took the context of offending into account and decided that risk *was* caused by the choice of a speed in excess of the posted limit. The human officer would, many drivers note, also be able to detect other dangerous road-user behaviours and would therefore further diffuse the objections that speed is overly targeted at the expense of other offences which the speed camera cannot detect. As such, the objections about speed enforcement targeting the 'wrong' places or the 'wrong' people would be lessened because such offences would not be targeted proactively. Such offences would technically remain illegal but could not be drawn on with such frequency as examples of illegitimate enforcement.

The use of the speed camera, however, exacerbates and magnifies these problematic aspects of speed enforcement by enforcing against *all* speeding drivers in *all* circumstances. The speed camera becomes the on-the-ground manifestation of the authorities' disinterest in the contextual aspects of an offence that determine whether or not an incident was genuinely risky. The inability of the detection methodology to assess the merits of particular cases and to focus on the demonstrably dangerous instances of speeding means that vast numbers of offences are produced that appear to have been generated despite the demonstrable 'safeness' of the act. Indeed, the issuing of a speeding ticket can be looked at, in this case, as proof of the absence of a harmful outcome, given that if harm had occurred other charges such as careless or dangerous driving would be used to legally problematise the behaviour. Objections can be deployed with enthusiasm, because the favoured enforcement method produces a vast number of 'unfair' ('safe') offences alongside the 'fair' ('unsafe') ones that it detects. A probabilistic risk issue is therefore pursued via a detection method that borders on a 'zero-tolerance' approach. As a result, speed cameras inevitably detect and punish instances of behaviour that fall outside (or are perceived to fall outside) the realms of probability. All the *potentially* problematic aspects of speed limit enforcement – the occasions when 'safe' behaviour can nonetheless be illegal – are rendered repeatedly *real* by the method of enforcement chosen. The speed camera techno-fix is therefore viewed as vital to the production and perpetuation of the debate around speed limit enforcement because it cannot offer the discretion or common sense that many drivers demand and which would be viewed as reflecting the probabilistic rather than deterministic nature of risk assessments made in relation to speed.

Therefore, in addition to providing the means through which implicated risk instigators are able to reject such a status, and the motive for them to engage in such a rejection, the use of 'risk' as motivation for enforcement also leads to the use of technologies that provide the specific opportunity for these concerns to surface and be deployed. The debate around the use of speed cameras as an enforcement technology can therefore be seen as a consequence of that same concern with risk that motivated their use in the first place.

Glossary

AA	Automobile Association
ABD	Association of British Drivers
ACPO	Association of Chief Police Officers
ANPR	Automatic Number Plate Recognition
DEFRA	Department of the Environment, Food and Rural Affairs
DETR	Department of the Environment, Transport and the Regions
DfT	Department for Transport
IAM	Institute of Advanced Motorists
KSI	Killed or Seriously Injured
MAD	Motorists Against Detection
MOT	Ministry of Transport test
NSCP	National Safety Camera Programme
PACTS	Parliamentary Advisory Council for Transport Safety
RAC	Royal Automobile Club
SAC	Speed Awareness Course
SPECS	Speed Enforcement Camera System
SID	Speed Indication Device
SSI	Slower Speeds Initiative
TLGR	Transport, Local Government and the Regions
TRL	Transport Research Laboratory
UCL	University College London
VAS	Vehicle Activated Sign

Bibliography

AA Motoring Trust 2002. *Road Traffic Speed: Memorandum Submitted by the Automobile Association to the Transport, Local Government and the Regions Committee.* London: AA Motoring Trust.

AA Motoring Trust 2003. *Intelligent Speed Adaptation.* London: AA Motoring Trust.

AA Motoring Trust 2005. Support drops for graduated penalties when speeding link to pedestrian injury is explained (press release, 8 September). London: AA Motoring Trust.

ABD 1999a. Does a 1mph reduction in speed *really* reduce accidents by 5%? Online, available at: www.abd.org.uk/onemph.htm (accessed 28 November 2003).

ABD 1999b. Fallacy of 'Speed Kills' exposed by authoritative report (press release, 3 July). London: Pro-Motor.

ABD 1999c. Response to the Government consultation on speed policy. Online, available at: www.abd.org.uk/spdconr.htm (accessed 20 April 2006).

ABD 1999d. Road safety – a bedtime story to give anybody a nightmare. Online, available at: www.abd.org.uk/nightmar.htm (accessed 9 October 2006).

ABD 2001. Rigging the evidence. Online, available at: www.abd.org.uk/rigging_the_evidence.htm (accessed 10 April 2005).

ABD 2002. One third = 7%: The biggest lie of all. Online, available at: www.abd.org.uk/one_third.htm (accessed 28 November 2003).

ABD 2003a. 'Speed cameras policy responsible for 5500 deaths' says the ABD (press release, 3 October). London: Pro-Motor.

ABD 2003b. ABD calls for an end to speed camera secrecy (press release, 30 November). London: Pro-Motor.

ABD 2004a. New speed camera rules are an extension of failing and dangerous policy (press release, 14 August). London: Pro-Motor.

ABD 2004b. ABD calls for end to speed camera spin (press release, 20 May). London: Pro-Motor.

ABD 2004c. Foundation stone of 'Speed Kills' abandoned by DfT (press release, 3 September). London: Pro-Motor.

ABD 2004d. Increased road death figures demolish Darling's camera claims (press release, 25 June). London: Pro-Motor.

ABD 2004e. Road safety group calls for independent speed camera investigation (press release, 28 July). London: Pro-Motor.

ABD 2004f. Shrines should save lives (press release, 2 November). London: Pro-Motor.

ABD 2005a. ABD predicts misuse of in car black boxes ... as TRL blames bad driving for higher road deaths (press release, 20 September). London: Pro-Motor.

ABD 2005b. J.J. Leeming – Accidental Expert. Online, available at: www.abd.org. uk/jjleeming.htm (accessed 28 March 2005).

ABD 2005c. ACPO in speed camera disarray after Brunstrom speaks out (press release, 16 March). London: Pro-Motor.

ABD 2006a. ABD condemns Stradling camera report as 'contrived' (press release, 27 April). London: Pro-Motor.

ABD 2006b. Richard Brunstrom: The Former Chief Constable of Controversy. Online, available at: www.abd.org.uk/brunstrom.htm (accessed 18 September 2006).

ABD 2008. Evidence to the Transport Committee's inquiry into road safety. Online, available at: www.abd.org.uk/resources/documents/transcom_roadsafety_2008.htm (accessed 18 October 2010).

ABD 2009. Join us! Have you had enough yet? Online, available at: www.abd.org. uk/about/join.htm (accessed 18 August 2010).

ABD 2010. Quotes on speed cameras. Online, available at: www.abd.org.uk/ resources/quotes/speed_cameras.htm (accessed 18 August 2010).

ACPO 2004. *Speed Enforcement Guidelines: Joining Forces for Safer Roads.* London: ACPO.

ACPO 2005. *ANPR Strategy for the Police Service 2005–8: Denying Criminals the Use of the Road.* London: ACPO.

ACPO 2006. *Guidance for the Introduction of a National Speed Awareness Course.* London: ACPO.

ACPO 2009. *National Driver Offender Re-Training Schemes – National Speed Awareness Course Guidance Notes.* London: ACPO.

ACPO, DfT and Home Office 2005. *Roads Policing Strategy.* London: ACPO/ DfT/Home Office.

ActTravelWise 2002. Speed camera challenge goes to the High Court. Online, available at: www.acttravelwise.org/news/13 (accessed 12 August 2010).

Adams, J. 2003. Risk and Morality: Three Framing Devices. In R. Ericson and A. Doyle (eds) *Risk and Morality.* London: University of Toronto Press, 87–103.

Aird, A. 1972. *The Automotive Nightmare.* London: Hutchinson.

Allsop, R. 2010. *The Effectiveness of Speed Cameras.* London: RAC Foundation.

Amadi, J. 2008. Piloting penalty notices for disorder on 10- to 15-year-olds: Results from a one year pilot. Online, Ministry of Justice Research Series 19/08, available at: www.justice.gov.uk/publications/docs/piloting-penalty-notices.pdf (accessed 13 May 2010).

Ashton, S. and MacKay, G. 1979. Some characteristics of the population who suffer trauma as pedestrians when hit by cars and some resulting implications, 4th IRCOBI International Conference, Gothenburg, 1979. Online, available at: http://webarchive.nationalarchives.gov.uk/+/http://www.dft.gov.uk/foi/ responses/2005/nov/203040message/paperaboutthedepartments20302445.

Atkinson, P. and Coffey, A. 2002. Analysing Documentary Realities. In D. Silverman (ed.) *Qualitative Research: Theory, Method and Practice*. London: Sage.

Autonational 2004. Survey's new solution to speeding. Online (press release), previously available at www.autonational.co.uk/news/2004_06.htm (accssed 10 September 2006).

Autonational Rescue 2009. Speed or safety in numbers. Online, available at: www.autonational.co.uk/2009_08.php (accessed 13 July 2010).

Baggott, R. 1995. *Pressure Groups Today*. Manchester: Manchester University Press.

Baker, T. 2003. Containing the Promise of Insurance: Adverse Selection and Risk Classification. In R. Ericson and A. Doyle (eds) *Risk and Morality*. London: University of Toronto Press, 87–103.

Barker, T. (ed.) 1987. *The Economic and Social Effects of the Spread of Motor Vehicles*. Basingstoke: Macmillan.

Bauman, Z. 1997. *Postmodernity and its Discontents*. Cambridge: Polity Press.

BBC Breakfast News 2003. Slow down for day of action. Online 10 December, available at: http://news.bbc.co.uk/1/hi/programmes/breakfast/3304093.stm (accessed 10 December 2003).

BBC Inside Out 2004. Speed – trading places. Online 20 September, available at: www.bbc.co.uk/insideout/northeast/series6/cameras.shtml (accessed 26 July 2010).

BBC News Online 2001. Fake speed camera signs spark row. Online 14 August, available at: http://news.bbc.co.uk/1/hi/uk/1490318.stm (accessed 17 March 2006).

BBC News Online 2002a. Cardboard camera tricks drivers. Online 2 September, available at: http://news.bbc.co.uk/1/hi/england/west_yorkshire/3200807.stm (accessed 17 March 2006).

BBC News Online 2002b. Arrests after number plate scans. Online 21 May, available at: http://news.bbc.co.uk/1/hi/england/3045807.stm (accessed 18 May 2006).

BBC News Online 2003a. 'Fake' speed camera fears. Online 23 November, available at: http://news.bbc.co.uk/1/hi/england/wiltshire/3229754.stm (accessed 17 March 2006).

BBC News Online 2003b. Speed camera stance defended. Online 1 September, available at: http://news.bbc.co.uk/1/hi/england/tees/3197995.stm (accessed 10 December 2003).

BBC News Online 2003c. Force in row over speed cameras. Online 10 December, available at: http://news.bbc.co.uk/1/hi/england/3305719.stm (accessed 10 December 2003).

BBC News Online 2003d. Speed cameras 'cause road deaths'. Online 5 December, available at: http://news.bbc.co.uk/1/hi/uk/3293611.stm (accessed 12 December 2003).

BBC News Online 2004a. Police chief's speed camera fears. Online 21 February, available at: http://news.bbc.co.uk/1/hi/uk/3509373.stm (accessed 21 February 2004).

BBC News Online 2004b. Unwritten rules of the motorway. Online 18 August, available at: http://news.bbc.co.uk/1/hi/magazine/3574334.stm (accessed 2 August 2006).

BBC News Online 2004c. Number plate scan nets criminals. Online 3 November, available at: http://news.bbc.co.uk/1/hi/england/lancashire/3976431.stm (accessed 18 May 2006).

BBC News Online 2005a. The unlikely traffic cops. Online 21 February, available at: http://news.bbc.co.uk/1/hi/magazine/4268873.stm (accessed 17 March 2006).

BBC News Online 2005b. Villagers given police radar guns. Online 7 January, available at: http://news.bbc.co.uk/1/hi/wales/4153429.stm (accessed 17 March 2006).

BBC News Online 2005c. Chief questions speed camera use. Online 7 November, available at: http://news.bbc.co.uk/1/hi/uk/4414370.stm (accessed 10 December 2003).

BBC News Online 2005d. Speed cameras target M4 drivers. Online 13 April 2005, available at: http://news.bbc.co.uk/1/hi/england/4439123.stm (accessed 15 April 2005).

BBC News Online 2005e. No new speed cameras until report. Online 15 July, available at: http://news.bbc.co.uk/1/hi/uk/4685337.stm (accessed 15 July 2005).

BBC News Online 2005f. Drive for safe roads near schools. Online 8 November, available at: http://news.bbc.co.uk/1/hi/scotland/4415716.stm (accessed 8 November 2005).

BBC News Online 2005g. Doctors attack 'smoke ban myths'. Online 27 August, available at: http://news.bbc.co.uk/1/hi/health/4489463.stm (accessed 13 September 2006).

BBC News Online 2005h. 16 held in number plate scan test. Online 2 February, available at: http://news.bbc.co.uk/1/hi/england/devon/4223173.stm (accessed 18 May 2006).

BBC News Online 2006a. Parishioners hire radar speed gun. Online 5 January, available at: http://news.bbc.co.uk/1/hi/england/devon/4584812.stm (accessed 26 April 2006).

BBC News Online 2006b. Crash risk is linked to speeding. Online 26 April, available at: http://news.bbc.co.uk/1/hi/england/west_midlands/4944834.stm (accessed 26 April 2006).

BBC News Online 2006c. Camera to focus on driver's face. Online 24 May, available at: http://news.bbc.co.uk/1/hi/uk/5012832.stm (accessed 29 May 2006).

BBC News Online 2006d. Man jailed for speed camera blast. Online 6 September, available at: http://news.bbc.co.uk/1/hi/england/manchester/5320092.stm (accessed 6 September 2006).

BBC News Online 2007a. CCTV to safeguard speed cameras. Online 24 January, available at: http://news.bbc.co.uk/1/hi/scotland/south_of_scotland/6293823. stm (accessed 18 February 2007).

BBC News Online 2007b. Family's plea for safer driving. Online 7 November, available at: www.bbc.co.uk/norfolk/content/articles/2007/11/05/feature_kai_ davies_20071105.shtml (accessed 20 July 2010).

BBC News Online 2008. Residents start speed gun project. Online 2 February, available at: http://news.bbc.co.uk/1/hi/england/nottinghamshire/7224073.stm (accessed 2 February 2010).

BBC News Online 2010a. Bird box resembling speed camera slows Wearside drivers. Online 23 February, available at: http://news.bbc.co.uk/1/hi/england/ cumbria/8529813.stm (accessed 23 February 2010).

BBC News Online 2010b. Speed camera cuts 'mean disaster'. Online 25 July, available at: www.bbc.co.uk/news/uk-10755509 (accessed 16 August 2010).

BBC News Online 2010c. Oxfordshire speeding increase after cameras turned off. Online 10 August, available at: www.bbc.co.uk/news/uk-england-oxfordshire-10929488 (accessed 11 August 2010).

BBC News Online 2010d. ACPO's Mick Giannasi fears speed camera cut risks lives. Online 9 August, available at: www.bbc.co.uk/news/uk-10911436 (accessed 11 August 2010).

BBC News Online 2010e. Is it the end of the road for the speed camera? Online 22 July 2010, available at: www.bbc.co.uk/news/uk-10723343 (accessed 11 August 2010).

BBC News Online 2011a. Oxfordshire speed cameras on again eight months after switch off. Online 1 April, available at: www.bbc.co.uk/news/uk-12931064 (accessed 1 April 2011).

BBC News Online 2011b. Fines planned for careless driving. Online, 11 May, available at www.bbc.co.uk/news/uk-13356057 (accessed 11 May 2011).

Beck, U. 1992. *Risk Society: Towards a New Modernity*. London: Sage.

Beck, U. 1998. Politics of Risk Society. In J. Franklin (ed.) *The Politics of Risk Society*. Cambridge: Polity Press.

Becker, H. 1963/1997. *Outsiders: Studies in the Sociology of Deviance*. New York: Free Press.

Blackspot.com 2010. Road Angel: Protecting your life and livelihood. Online, available at: www.blackspot.com (accessed 23 August 2011).

Blincoe, K. et al. 2006. Speeding drivers' attitudes and perceptions of speed cameras in rural England. *Accident Analysis and Prevention*, 38(2), 371–8.

Bloor, M. et al. 2001. *Focus Groups in Social Research*. London: Sage.

Booker, C. and North, R. 2007. *Scared to Death: The Anatomy of a Very Dangerous Phenomenon*. London: Continuum.

Brake 2004. Speed. Online, available at: www.brake.org.uk/assets/docs/Factsandresources/Slow_Up_2004.pdf (accessed 28 September 2004).

Broughton, J., Markey, K. and Rowe, D. 1998. *A New System for Recording Contributory Factors in Road Accidents: TRL 323*. Crowthorne: Transport Research Laboratory.

Brunner, C.T. 1928. *The Problem of Motor Transport*. London: Ernest Benn Ltd.

Buckingham, A. 2003. Speed traps: Saving lives or raising revenue? *Policy*, 19(3), 3–12.

Cameron, M. and Buckingham A. 2003. Speed off. *Policy*, 19(4), 60–64.

Campbell, D. and Evans, R. 2006. Surveillance on drivers may be increased. *The Guardian*, 7 March, 1.

Campbell, M. and Stradling, S. 2003. *Factors Influencing Driver Speed Choices: Road Safety Research Report No. 19*. London: DfT.

Captain Gatso 2002. Fleecing, not policing! Online: Motorists Against Detection (press release, 22 August), available at: www.speedcam.co.uk/index2.htm (accessed 28 February 2002).

Carroll, A. 2002. *Speeding Excuses that Work*. California: Gray Area Press.

Chan, J. 2003. Police and New Technologies. In T. Newburn (ed.) *Handbook of Policing*. Devon: Willan.

Clayton, S. and Opotow, S. 2003. Justice and identity: Changing perspectives on what is fair. *Personality and Social Psychology Review*, 7(4), 298–310.

Coleman, R. 2004. *Reclaiming the Streets*. Devon: Willan.

Corbett, C. 1995. Road traffic offending and the introduction of speed cameras in England: The first self-report study. *Accident Analysis and Prevention*, 27(3), 345–54.

Corbett, C. 2000. A typology of drivers' responses to speed cameras: Implications for speed limit enforcement and road safety. *Psychology, Crime and Law*, 6(4), 1–26.

Corbett, C. 2003. *Car Crime*. Devon: Willan.

Corbett, C. 2008. Roads policing: Current context and imminent dangers. *Policing: A Journal of Policy and Practice*, 2(1), 131–42.

Corbett, C. and Caramlau, I. 2006. Gender differences in responses to speed cameras: Typology findings and implications for road safety. *Criminology and Criminal Justice*, 4, 411–33.

Corbett, C. and Grayson, G. 2010. Speed limit enforcement as perceived by offenders: Implications for roads policing. *Policing: A Journal of Policy and Practice*, 4(2), 364–72.

Corbett, C. and Simon, F. 1991. Police and public perceptions of the seriousness of traffic offences. *British Journal of Criminology*, 31(2), 153–64.

Corbett, C. and Simon, F. 1992. Decisions to break or adhere to the rules of the road, viewed from the rational choice perspective. *British Journal of Criminology*, 32(4), 537–49.

Corbett, C. and Simon, F. 1999. *The Effects of Speed Cameras: How Drivers Respond: Road Safety Research Report No. 11*. London: DETR.

Corbett, C. et al. 2008. *Does the Threat of Disqualification Deter Drivers from Speeding?* London: DfT.

Cornick, D. 1997. Cyberspace: Its Impact on the Conventional Way of Doing and Thinking About Research. Sixth Annual Conference of the Urban Business Association, 29 April 1995. Previously available at: www.csaf.org/cyber.htm (accessed 24 June 2006).

Daily Express 2001. A killer we all ignore. 31 March, 30.

Daily Express 2002. Police profits before lives. 20 May, 1.

Daily Express 2010a. Outrage at speed camera that rakes in £1.3m a year on safe road. Online 8 July, available at: www.express.co.uk/posts/view/185580/Outrage-at-speed-camera-that-rakes-in-1-3m-a-year-on-safe-road (accessed 8 July 2010).

Daily Express 2010b. Speed cameras may have caused 28,000 crashes. Online 6 August, available at: www.express.co.uk/posts/view/191443/Speed-cameras-may-have-caused-28-000-crashes (accessed 11 August 2010).

Daily Mail 2003. Fall of the cameras. 31 October, 43.

Daily Mail 2004. Speed camera police rake in £5,100 a year extra wages. 30 January, 29.

Daily Mail 2005. The lying camera. 24 December, 7.

Daily Mail 2006a. Speed camera report 'absurd', says road safety group. Online 22 November, available at: www.dailymail.co.uk/news/article-417928/Speed-camera-report-absurd-says-road-safety-group.html# (accessed 18 August 2010).

Daily Mail 2006b. Stubbed out. 15 February, 7.

Daily Mail 2009. England's most lucrative speed camera revealed – Raking in £2.3million in five years. Online 28 December, available at: www.dailymail.co.uk/news/article-1238885/Englands-lucrative-speed-camera-revealed-raking-2-3million-years.html (accessed 4 January 2010).

Daily Mail 2010a. Duke of Edinburgh avoids trouble with a speed camera detector. Online 27 February, available at: www.dailymail.co.uk/news/article-1254170/Caught-Duke-Edinburgh-aims-beat-the.html (accessed 22 July 2010).

Daily Mail 2010b. Ministers promise to end funding for speed camera 'cash-cow'. Online 18 June, available at: www.dailymail.co.uk/news/article-1287496/Ministers-promise-end-funding-speed-camera-cash-cow.html (accessed 18 June 2010).

Daily Telegraph 2001. Motormouth. 8 September, 5.

Daily Telegraph 2003a. Driver with circular saw snaps speed camera. Online 20 May, available at: www.telegraph.co.uk/news/main.jhtml?xml=/news/2003/05/20/nsaw20.xml (accessed 28 May 2004).

Daily Telegraph 2003b. £4000 fine over speed camera revenge. 6 June, 3.

Daily Telegraph 2004. Assaults are up – on the poor old British motorist. Online 13 January, available at: www.telegraph.co.uk/comment/telegraph-view/3601415/Assaults-are-up-on-the-poor-old-British-motorist.html (accessed 13 January 2004).

Daily Telegraph 2005a. Drive ban over speed camera fire. 7 October, 1.

Daily Telegraph 2005b. Technology to drive revolution on road safety. Online 19 October, available at: www.telegraph.co.uk/news/uknews/1500301/Technology-to-drive-revolution-in-road-safety.html (accessed 10 October 2005).

Daily Telegraph 2005c. Six traffic areas clock up £1m-plus speed profits. Online 27 December, available at: www.telegraph.co.uk/news/main.jhtml?xml=/news/2005/12/27/nspeed27.xml (accessed 3 May 2006).

Daily Telegraph 2006. Smoking ban leaves a fog of confusion. 15 February, 6.

Daily Telegraph 2010a. Traffic camera rakes in nearly £1 million in a month. Online 13 May, available at: www.telegraph.co.uk/news/uknews/road-and-rail-transport/7718344/Traffic-camera-rakes-in-nearly-1-million-in-a-month.html (accessed 5 July 2010).

Daily Telegraph 2010b. RAC: local authorities cutting back on speed cameras. Online 18 June, available at: www.telegraph.co.uk/motoring/news/7838300/RAC-local-authorities-cutting-back-on-speed-cameras.html (accessed 18 June 2010).

Daily Telegraph 2010c. Treasury set to cash in on speeding fines. Online 26 July, available at: www.telegraph.co.uk/motoring/news/7909246/Treasury-set-to-cash-in-on-speeding-fines.html (accessed 11 August 2010).

Daily Telegraph 2010d. Average speed cameras to be installed on major urban stretch for first time. Online 18 January, available at: www.telegraph.co.uk/motoring/news/7020694/Average-speed-cameras-to-be-installed-on-major-urban-stretch-for-first-time.html (accessed 20 August 2010).

Daily Telegraph 2010e. Speed camera switch-off sees fewer accidents. Online 7 August, available at: www.telegraph.co.uk/motoring/news/7931842/Speed-camera-switch-off-sees-fewer-accidents.html (accessed 11 August 2010).

Daily Telegraph 2010f. Safety groups launch bid to save speed cameras. Online 31 August, available at: www.telegraph.co.uk/motoring/road-safety/7973246/Safety-groups-launch-bid-to-save-speed-cameras.html (accessed 31 August 2010).

Datamonitor 2006. Speed cameras and speeding drivers. Online: Safespeed, available at: www.safespeed.org.uk/swiftcover2006.doc (accessed 11 April 2006).

Davis, G. 2002. Is the claim that 'speed variance kills' an ecological fallacy? *Accident Analysis and Prevention*, 34(3), 343–6.

DEFRA 2005. Clean Neighbourhoods and Environment Act – Outline of Measures. Online: Department for Environment, Food and Rural Affairs, available at: www.defra.gov.uk/environment/quality/local/legislation/cnea/documents/leqbill-summary.pdf (accessed 20 September 2006).

Deleuze, G. 1995. *Negotiations 1972–1990*. New York: Columbia University Press.

Delhomme, P. 1991. Comparing one's driving with others: Assessment of abilities and frequency of offences. Evidence for a superior conformity or self bias? *Accident Analysis and Prevention*, 23, 493–508.

DETR 1999a. Speed cameras funding – Eight pilot schemes announced. Online 16 December, available at: http://trid.trb.org/view.aspx?id=651953 (accessed 23 August 2011).

DETR 1999b. *Speed Policy Review*. London: DETR.

DETR 2000a. *Tomorrow's Roads: Safer for Everyone*. London: DETR.

DETR 2000b. *New Directions in Speed Management: A Review of Policy*. London: DETR.

DETR 2001. Spellar announces new camera visibility rules (DTLR press release 517). London: DTLR.

DfT 1991. *Road Traffic Act 1991*. London: HMSO.

DfT 2002. *Behavioural Research in Road Safety: Eleventh Seminar Proceedings*. London: DfT Publications.

DfT 2003a. *A Cost Recovery System for Speed and Red-Light Cameras – Two Year Pilot Evaluation*. London: DfT Publications.

DfT 2003b. *Vehicle Speeds in Great Britain: 2002*. London: DfT Publications.

DfT 2003c. *Behavioural Research in Road Safety: Twelfth Seminar Proceedings*. London: DfT Publications.

DfT 2004a. *Behavioural Research in Road Safety: Fourteenth Seminar Proceedings*. London: DfT Publications.

DfT 2004b. Graduated fixed penalties for speeding offences – Discussion note. Online: Department for Transport, available at: http://webarchive. nationalarchives.gov.uk/+/http://www.dft.gov.uk/pgr/roadsafety/ speedmanagement/graduatedfixedpenaltiesforsp4801 (accessed 23 August 2011).

DfT 2004c. Graduated speeding penalties: Consultation launched (press release, 1 September). London: DfT.

DfT 2004d. *Handbook of Rules and Guidance for the National Safety Camera Programme for England and Wales for 2005/6*. London: DfT Publications.

DfT 2005a. Road Safety Bill Regulatory Impact Assessment. Online: Department for Transport, available at: http://webarchive.nationalarchives.gov. uk/20050301192906/http:/dft.gov.uk/stellent/groups/dft_rdsafety/documents/ pdf/dft_rdsafety_pdf_033069.pdf (accessed 23 August 2011).

DfT 2005b. Greater funding and flexibility for road safety as report shows cameras working (press release, 15 December). London: DfT.

DfT 2005c. *Behavioural Research in Road Safety: Fifteenth Seminar Proceedings* London: DfT Publications.

DfT 2005d. Latest speed campaign. Think! Online: Department for Transport, previously available at: www.thinkroadsafety.gov.uk/campaigns/slowdown/ slowdown.htm (accessed 20 June 2006).

DfT 2005e. Think! Slow down – Previous campaign. Online: Department for Transport, previously available at: www.thinkroadsafety.gov.uk/campaigns/slowdown/previous.htm (accessed 20 June 2006).

DfT 2005f. Road Safety Bill 2005 – Briefing notes. Local Transport Planning Network. Online: Department for Transport, available at: www.ltpnetwork.gov.uk/Documents/Document/road%20safety%20bill%20summary.pdf (accessed 21 April 2006).

DfT 2006a. *Handbook of Rules and Guidance for the National Safety Camera Programme for England and Wales for 2006/7*. London: DfT Publications.

DfT 2006b. *Setting Local Speed Limits* (DfT circular, January). London: HMSO.

DfT 2006c. *Behavioural Research in Road Safety: Sixteenth Seminar Proceedings*. London: DfT Publications.

DfT 2006d. Road Safety Strategy Division Research Programme – Summary of planned research. Online: Department for Transport, previously available at: http://www.dft.gov.uk/stellent/groups/dft_rdsafety/documents/pdf/dft_rdsafety_pdf_611265.pdf (accessed 18 June 2006).

DfT 2007. *Behavioural Research in Road Safety: Seventeenth Seminar Proceedings* London: DfT Publications.

DfT 2008. Road Safety Compliance Consultation. Online: Department for Transport, November, available at: www.dft.gov.uk/consultations/closed/compliance/roadsafetyconsultation.pdf (accessed 13 May 2009).

DfT 2009a. Frequently asked questions: What is the history of speed cameras? Online, available at: www.dft.gov.uk/pgr/roadsafety/speedmanagement/safety camerasfrequentlyasked4603?page=1#a1014 (accessed 18 May 2010).

DfT 2009b. *A Safer Way: Consultation Document on Making Britain's Roads the Safest in the World*. London: DfT Publications.

DfT 2009c. Chilling campaign warns drivers to kill their speed or live with it. Online: Department for Transport (news release 016), 30 January, available at: www.pacts.org.uk/news.php?id=285 (accessed 23 August 2011).

DfT 2009d. *Road User Safety Division: Call for Comments on Revision of DfT's Speed Limit (Circular 01/06)*. London: DfT Publications.

DfT 2010. Campaign warns drivers of the consequences of speeding. Online: Department for Transport (news release, 4 January 2010), available at: www.driving-news.co.uk/driving-legislation/campaign-warns-drivers-of-the-consequences-of-speeding (accessed 23 August 2011).

DfT 2011. *Strategic Framework for Road Safety*. London: DfT Publications.

Druckman, J.N. and Nelson, K.R. 2003. Framing and deliberation: How citizens' conversations limit elite influence. *American Journal of Political Science*, 47(4), 729–45.

Duff, R.A. 2002. Crime, prohibition and punishment. *Journal of Applied Philosophy*, 19(2), 97–108.

Emsley, C. 1993. Mother, what *did* policemen do when there weren't any motors? The law, the police and the regulation of motor traffic in England, 1900–1939. *The Historical Journal*, 30(2), 357–81.

Ericson, R. 1994. The division of expert knowledge in policing and security. *British Journal of Sociology*, 45(2), 149–75.

Ericson, R. and Doyle, A. (eds) 2003. *Risk and Morality*. London: University of Toronto Press.

Ericson, R. and Haggerty, K. 1997. *Policing the Risk Society*. Oxford: Clarendon.

Fleiter, J., Lennon, A. and Watson, B. 2007. *Choosing Not to Speed: A Qualitative Exploration of Differences in Perceptions about Speed Limit Compliance and Related Issues*. Australia: Centre for Accident Research and Road Safety.

Flower, R. and Wynn-Jones, M. 1981. *One Hundred Years of Motoring: An RAC Social History of the Car*. Maidenhead: McGraw-Hill.

Foreman-Peck, J. 1987. Death on the Roads: Changing National Responses to Motor Accidents. In T. Barker (ed.) *The Economic and Social Effects of the Spread of Motor Vehicles*. Basingstoke: MacMillan.

Fox, R. 1995. *Criminal Justice on the Spot: Infringement Penalties in Victoria*. Canberra: Australian Institute of Criminology.

Franklin, J. (ed.) 1998. *The Politics of Risk Society*. Cambridge: Polity Press.

Gabor, T. 1994. *Everybody Does It! Crime by the Public*. Toronto: University of Toronto Press.

Gains, A. et al. 2004. *The National Safety Camera Programme – Three-Year Evaluation Report*. London: PA Consulting.

Gains, A. et al. 2005. *The National Safety Camera Programme – Four-Year Evaluation Report*. London: PA Consulting.

Garland, D. 1996. The limits of the sovereign state. *British Journal of Criminology*, 36(4), 445–71.

Garland, D. 2003. The Rise of Risk. In R. Ericson and A. Doyle (eds) *Risk and Morality*. London: University of Toronto Press, 48–86.

Garland, D. and Sparks, R. 2000. Criminology, Social Theory and the Challenge of Our Times. In D. Garland and R. Sparks (eds) *Criminology and Social Theory*. Oxford: Clarendon, 1–22.

Geary, R. 1994. *Essential Criminal Law*. London: Cavendish.

Giddens, A. 1991. *Modernity and Self-Identity*. Cambridge: Polity Press.

Giddens, A. 1998. Risk Society: The Context of British Politics. In J. Franklin (ed.) *The Politics of Risk Society*. Cambridge: Polity Press, 23–34.

Giddens, A. 1999. Risk and responsibility. *Modern Law Review*, 62, 1–10.

Gifford, R. 2004. Why reducing your speed does matter. *The Independent*. Online 7 September, available at: www.independent.co.uk/life-style/motoring/comment/robert-gifford-why-reducing-your-speed-does-matter-542734.html (accessed 23 August 2011).

Girling, E., Loader, I. and Sparks, R. 2000. *Crime and Social Change in Middle England*. London: Routledge.

Goffman, E. 1959/2010. *The Presentation of Self in Everyday Life*. London: Penguin.

Grayson, G.B. (ed.) 1997. *Behavioural Research in Road Safety VII*. Crowthorne: Transport Research Laboratory.

Groeger, J.A. and Brown I.D. 1989. Assessing one's own and others' driving ability: Influences of sex, age and experience. *Accident Analysis and Prevention*, 21(2), 155–68.

The Guardian 2004a. They call themselves the voice of the driver. But who do they really represent? Online 3 February, available at: www.guardian.co.uk/uk/2004/feb/03/transport.world (accessed 23 August 2011).

The Guardian 2004b. Another fine mess. Online 14 January, available at: www.guardian.co.uk/politics/2004/jan/14/immigrationpolicy.prisonsandprobation1 (accessed 23 August 2011).

The Guardian 2006. Why speed cameras need safety cameras. 11 July, 2–3.

Gusfield, J.R. 1981. *The Culture of Public Problems*. London: University of Chicago Press.

Hacking, I. 2003. Risk and Dirt. In R. Ericson and A. Doyle (eds) *Risk and Morality*. London: University of Toronto Press, 22–47.

Haggerty, K. 2002. The politics of statistics: Variations on a theme. *Canadian Journal of Sociology*, 27(1), 89–105.

Haggerty, K. 2003. From Risk to Precaution: The Rationalities of Personal Crime Prevention. In R. Ericson and A. Doyle (eds) *Risk and Morality*. London: University of Toronto Press, 193–215.

Haggerty, K. 2004a. Technology and crime policy – Reply to Michael Jacobson. *Theoretical Criminology*, 8(4), 491–7.

Haggerty, K. 2004b. Displaced expertise: Three constraints on the policy-relevance of criminological thought. *Theoretical Criminology*, 8(2), 211–31.

Hansard HL 1995, 30 October. *Dogs (Fouling of Land) Bill, Vol. 566, Part No. 135, Col. 1292*.

Hebenton, B. and Seddon, T. 2009. From dangerousness to precaution: Managing sexual and violent offenders in an insecure and uncertain age. *British Journal of Criminology*, 49(3), 343–62.

Heimer, C. 2003. Insurers as Moral Actors. In R. Ericson and A. Doyle (eds) *Risk and Morality*. London: University of Toronto Press, 284–316.

Hennessy, D.A. and Wiesenthal, D.L. (eds) 2005. *Current Trends in Driver Behavior and Traffic Safety Research*. New York: Nova Science Publishers.

Hine, C. 2000. *Virtual Ethnography*. London: Sage.

Hirst, W.M., Mountain, L.J. and Maher, M.J. 2005. Are speed enforcement cameras more effective than other speed management measures? An evaluation of the relationship between speed and accident reductions. *Accident Analysis and Prevention*, 37(4), 731–41.

Holstein, J.A. and Gubrium, J.F. 2002. Active Interviewing. In D. Silverman (ed.) *Qualitative Research: Theory, Method and Practice*. London: Sage.

Home Office 2000. *Reducing Public Disorder: The Role of Fixed Penalty Notices: A Consultation Paper*. London: Home Office.

Home Office 2006. *Summary of Responses to the Consultation Document Standard Powers for Community Support Officers and a Framework for the Future Development of Power and the Government Response*. London: HMSO.

Home Office, ACPO and PA Consulting 2004. *Driving Crime Down: Denying Criminals the Use of the Road*. London: HMSO.

Hooke, A., Knox, J. and Portas, D. 1996. *Cost Benefit Analysis of Traffic Light and Speed Cameras* (Police Research Series, Paper 20). London: Home Office Police Research Group.

House of Commons Transport Select Committee 2004. *Traffic Law and its Enforcement: Sixteenth Report of Session 2003/4*. London: HMSO

Hudson, B. 2001. Punishment, Rights and Difference: Defending Justice in the Risk Society. In K. Stenson and R. Sullivan (eds) *Crime, Risk and Justice*. Devon: Willan, 144–72.

Hudson, B. 2003. *Justice in the Risk Society*. London: Sage.

Hunt, A. 2003. Risk and Moralization in Everyday Life. In R. Ericson and A. Doyle (eds) *Risk and Morality*. London: University of Toronto Press, 165–92.

Illingworth, N. 2001. The internet matters: Exploring the use of the internet as a research tool. *Sociological Research Online*, 6(2), available at: www.socresonline.org.uk/6/2/illingworth.html (accessed 23 August 2011).

The Independent 2010a. Speed camera 'cash cow' dries up. Online 17 June, available at: www.independent.co.uk/news/uk/home-news/speed-camera-cash-cow-dries-up-2003115.html (accessed 17 June 2010).

The Independent 2010b. The cuts victim that won't provoke any tears: speed cameras. Online 26 July, available at: www.independent.co.uk/life-style/motoring/motoring-news/the-cuts-victim-that-wont-provoke-any-tears-speed-cameras-2035411.html (accessed 26 July 2010).

Institute of Advanced Motorists 2009. Young motorists most supportive of safety cameras says IAM. Online, available at: www.iam.org.uk/latest_news/youngmotoristsmostsupportiveofsafetycamerassaysiam.html (accessed 13 July 2010).

Jamieson, A. 1988. Social Movements and the Politicization of Science. In J. Annerstedt and A. Jamieson (eds) *From Research Policy to Social Intelligence*. Macmillan: Basgingstoke.

Joh, E. 2007. Discretionless policing: Technology and the Fourth Amendment. *California Law Review*, 95, 199–234.

Johnson, R. 2004. Citizen expectations of police traffic stop behavior. *Policing: An International Journal of Police Strategies and Management*, 27, 487–97.

Johnston, L. 2000. *Policing Britain: Risk, Security and Governance*. Essex: Pearson.

Keenan, D. 2002. Speed cameras: The true effect on behaviour. *Traffic Engineering and Control*, 43(4), 154–9.

Keenan, D. 2004. Speed cameras: How do drivers respond? *Traffic Engineering and Control*, 45(3), 104–11.

Klare, H.J. (ed.) 1966. *Changing Concepts of Crime and its Treatment*. Oxford: Penguin.

Kleinig, J. 1996. *The Ethics of Policing*. Cambridge: Cambridge University Press.

Lash, S. and Wynne, B. 1992. Introduction. In U. Beck (ed.) *Risk Society: Towards a New Modernity*. London: Sage, 1–16.

Lea, J. 2002. *Crime and Modernity*. London: Sage.

Leeming, J.J. 1969. *Road Accidents*. London: Cassell.

Leps, M.C. 1992. *Apprehending the Criminal: The Production of Deviance in Nineteenth-Century Discourse*. London: Duke University Press.

Levi, M. 1997. *Consent, Dissent and Patriotism*. New York: Cambridge University Press.

Lewis, S. 2005. *The Speedmeter Handbook (Fourth Edition): A Guide to Type Approval Procedures for Speedmeters Used for Road Traffic Law Enforcement in Great Britain*. St Albans: Home Office Scientific Development Branch.

Lianos, M. and Douglas, M. 2000. Dangerization and the End of Deviance: The Institutional Environment. In D. Garland and R. Sparks (eds) *Criminology and Social Theory*. Oxford: Clarendon, 103–26.

Loader, I. and Mulcahy, A. 2003. *Policing and the Condition of England*. Oxford: Clarendon.

London Accident Analysis Unit 1997/2003. West London Speed Camera Demonstration Project. Online, available at: www2.dft.gov.uk/pgr/roadsafety/speedmanagement/nscp/nscp/westlondonspeedcamerademonst4601.html (accessed 23 August 2011).

Luban, D. 2002. The publicity of law and the regulatory state. *Journal of Political Philosophy*, 10(3), 296–316.

McConville, M. and Wilson, G. (eds) 2002. *The Handbook of the Criminal Justice Process*. Oxford: Oxford University Press.

MacCoun, R.J. 2005. Voice, control and belonging: The double-edged sword of procedural fairness. *Annual Review of Law and Social Science*, 1, 171–201.

McKenna, F. 1993. 'It won't happen to me': Unrealistic optimism and the illusion of control. *British Journal of Psychology*, 84(1), 39–50.

McKenna, F. 2007. *Do Attitudes and Intentions Change across a Speed Awareness Workshop? Behavioural Research in Road Safety No. 17*. London: DfT.

McKenna, F. and Poulter, D. 2008. Speed Awareness: The Effect of Education Versus Punishment on Driver Attitudes. In W.R. Nickel and M. Korán (eds) *Fit to Drive 2008*. Bonn: Kirschbaum-Verlag.

McManus, F. 2000. Noise law in the United Kingdom: A very British solution? *Legal Studies*, 20(2), 264–90.

Magistrates' Association 2010. Reply to the Sentencing Policy and Practice Committee: The use of fines. Online: Magistrates' Association Policy 10/35, available at: www.magistrates-association.org.uk/dox/consultations/1274599976_use_of_fines_policy_11_may_2010.pdf?PHPSESSID=oop70vtgdm9dmhacc4hqn52p11 (accessed 27 April 2011).

Mail on Sunday 2004. 'I'm the police chief who brought speed cameras to Britain … but even I think they have gone too far now'. 1 Februrary, 15.

Maxwell, S. 2004. The Prohibition of Smoking in Regulated Areas (Scotland) Bill: Supplementary evidence received by the Health Committee, Session 2. Online

13 May 2004, available at: www.scottish.parliament.uk/business/committees/ health/inquiries/ros/he04-smo-326.htm (accessed 20 October 2006).

Miller, J. and Glassner, B. 2002. The 'Inside' and the 'Outside': Finding Realities in Interviews. In D. Silverman (ed.) *Qualitative Research: Theory, Method and Practice*. London: Sage, 99–112.

Millman, R. 2006. Aussie drivers in internet switch scam. Online: *Secure Business Intelligence Magazine*, 17 July, available at: www.scmagazineus.com/aussie-drivers-internet-switch-scam/article/33677 (accessed 23 August 2011).

Monbiot, G. 2005. Paul Smith and safe speed – The self-exposure of a crank. Online: George Monbiot, available at: www.monbiot.com/archives/2005/12/22/paul-smith-and-safe-speed-the-self-exposure-of-a-crank (accessed 11 January 2005).

Mori 2001. Direct Line survey reveals drivers approve of speed cameras. Online: Mori press release, 2 August, available at: www.ipsos-mori.com/ researchpublications/researcharchive/poll.aspx?oItemId=1202 (accessed 23 August 2011).

Morris, T. 1966. The Social Toleration of Crime. In H.J. KIare (ed.) *Changing Concepts of Crime and its Treatment*. Oxford: Penguin, 13–24.

MAD 2006. A Summer of MADness? Online: Motorists Against Detection press release, available at: www.speedcam.co.uk/index2.htm (accessed 8 August 2006).

Mountain, L. 2008. Speed cameras: 'They reduce accidents as well as raising revenue'. *The Guardian*. Online 18 September, available at: www.guardian. co.uk/science/audio/2008/sep/16/speed.cameras.linda.mountain (accessed 8 July 2010).

Musselwhite, C., Avinerir, D., Fulcher, E., Goodwin, P. and Susilo, Y. 2010. Public attitudes to road user safety: A review of the literature, 2000–2009, Universities Transport Studies Group Conference, University of Plymouth, January 2010. Online, available at: http://eprints.uwe.ac.uk/13367/ (accessed 13 July 2010).

Mutch, I. 2002. The great speed debate: The nation awakens. Online: StreetBiker, 124, previously available at: www.streetbiker-mag.com/sb0124/6.html (accessed 18 March 2004).

Mythen, G. 2005. Employment, individualisation and insecurity: Rethinking the risk society perspective. *The Sociological Review*, 8(1), 129–49.

National Statistics 2002. *Road Accident Casualties by Road User Type and Severity 1992–2002: Annual Abstract of Statistics*. London: Transport Statistics/DfT.

National Statistics 2003. *Attitudes to Road Safety*. London: Office for National Statistics.

National Statistics 2005. *National Travel Survey*. London: Transport Statistics/ DfT.

Newburn T. (ed.) 2003. *Handbook of Policing*. Devon: Willan.

North, P. 1998. *Road Traffic Law Review Report*. London: HMSO.

O'Connell, S. 1998. *The Car and British Society*. Manchester: Manchester University Press.

O'Malley, P. 2004. The uncertain promise of risk. *Australian and New Zealand Journal of Criminology*, 37, 323–42.

O'Malley, P. 2008. Experiments in risk and criminal justice. *Theoretical Criminology*, 12, 451–69.

O'Malley, P. 2009a. *The Currency of Justice: Fines and Damages in Consumer Societies*. London: Routledge.

O'Malley, P. 2009b. Theorizing fines. *Punishment and Society*, 11, 67–83.

O'Malley, P. 2010. Simulated justice: Risk, money and telemetric policing. *British Journal of Criminology*, 50, 795–807.

PACTS 2000. *'Tomorrow's Roads – Safer for Everyone': A Response from the Parliamentary Advisory Council for Transport Safety*. London: Parliamentary Advisory Council for Transport Safety.

PACTS 2005. *Policing Road Risk: Enforcement, Technologies and Road Safety*. London: Parliamentary Advisory Council for Transport Safety.

PACTS 2008. *Managing Speed: Towards Safe and Sustainable Road Transport: Report from the European Transport Safety Council*. London: Parliamentary Advisory Council for Transport Safety.

PACTS and SSI 2003. *Speed Cameras: 10 Criticisms and Why They Are Flawed*. London: Parliamentary Advisory Council for Transport Safety /SSI.

Parker, D. and Stradling, S. 2001. *Influencing Driver Attitudes and Behaviour: Road Safety Research Report, No. 17*. London: DETR.

Pilkington, P. and Kinra, S. 2005. Effectiveness of speed cameras in preventing road traffic collisions and related casualties: Systematic review. *British Medical Journal*, 330, 331–4.

Plowden, W. 1971. *The Motor Car and Politics in Britain*. Middlesex: Penguin.

Plowden, S. and Hillman, M. 1996. *Speed Control and Transport Policy*. London: Policy Studies Institute.

Povey, D. et al. 2010. *HOSB Police Powers and Procedures, England and Wales 2008/09*. London: Home Office.

ProtectionInsurance.com 2004. Speed camera insurance. Online, available at www.pressbox.co.uk/Detailed/14213.html (accessed 23 August 2011).

RAC Foundation 2002. *Motoring Towards 2050*. London: RAC Foundation.

RAC Foundation 2003. Education not disqualification: Talking sense on speed campaign (press release, 20 October). London: RAC Foundation.

RAC Foundation 2004. Social exclusion: Transport links explored (press release, 23 February). London: RAC Foundation.

RAC Foundation 2005. Speeding. Online, previously available at: www.racfoundation.org/index.php?option=com_content&task=view&id=56&Itemid=31 (accessed 13 June 2005).

RAC 2005. *The Agony and Ecstasy of Driving: RAC Report on Motoring – Summary*. London: RAC Motoring Services.

Raine, J., Dunstan, E. and Mackie, A. 2003. Financial penalties as a sentence of the court: Lessons for policy and practice from research in the magistrates' courts of England and Wales. *Criminology and Criminal Justice*, 3(2), 181.

RoadPeace 2003. Draft response to Transport Committee Inquiry: Traffic law and its enforcement. Online, available at: http://admin.roadpeace.org/projcamp/ traffresp.html (accessed 29 September 2006).

Roberts, P. 2002. Science, Experts and Criminal Justice. In M. McConville and G. Wilson (eds) *The Handbook of the Criminal Justice Process*. Oxford: Oxford University Press, 253–84.

Rose, N. 2000. Government and Control. In D. Garland and R. Sparks (eds) *Criminology and Social Theory*. Oxford: Clarendon, 183–208.

Ross, H.L. 1960. Traffic law violation: A folk crime. *Social Problems*, 8, 231–41.

Ross, H.L. 1973. Folk crime revisited. *Criminology*, 11(1), 71–86.

Rothengatter, T. and Huguenin, R. (eds) 2004. *Traffic and Transport Psychology: Theory and Application Vol. 2000*. Oxford: Elsevier.

Rutherford, M. 2001. Motormouth. *The Daily Telegraph,* 8 September, 5

Sadler, P. 2008. *Effectiveness of Speed Indicator Devices on Reducing Vehicle Speeds in London: London Road Safety Unit, Research Summary No. 13*. London: TfL.

St Christopher.com 2010. Driverguard Plus. Online, available at: www.st-christopher.com/policy_plus.asp (accessed 24 August 2011).

Scottish Parliament 2005. *Environmental Levy on Plastic Bags (Scotland) – Policy Memorandum*. Norwich: HMSO.

Secretary of State for Transport 2002. *The Government's Response to the Transport, Local Government and the Regions Committee's Report: Road Traffic Speed*. London: HMSO.

Seddon, T. 2004. Searching for the next techno-fix: Drug testing in the criminal justice system. *Criminal Justice Matters*, 58, 16–17.

Shephard Engel, R. 2005. Citizen's perceptions of distributive and procedural injustice during traffic stops with police. *Journal of Research in Crime and Delinquency*, 42, 445–81.

Slower Speeds Initiative n.d. Improve road safety. Online, available at: www. slower-speeds.org.uk/safety (accessed 18 August 2006).

Slower Speeds Initiative 2001. How dangerous is speed? The ABD's lonely 'factoid' and the real world. *Slower Speeds Initiative Newsletter*, Spring 2001.

Smith, P. 2003a. Talking sense on speed campaign. Online: Safe Speed, available at: www.safespeed.org.uk/tsos.html (accessed 20 October 2004).

Smith, P. 2003b. Safe Speed views receive academic validation. Online: Safe Speed, available at: www.safespeed.org.uk/buckingham.html (accessed 2 February 2004).

Smith, P. 2004a. Regression to the mean. Online: Safe Speed, available at: www. safespeed.org.uk/rttm.html (accessed 19 August 2010).

Smith, P. 2004b. Speed cameras – The case against. Online: Safe Speed, available at: www.safespeed.org.uk/againstcameras.doc (accessed 20 October 2004).

Smith, P. 2006a. Welcome to Safe Speed: Dedicated to intelligent road safety. Online: Safe Speed, available at: www.safespeed.org.uk/main.html (accessed 18 August 2006).

Smith, P. 2006b. Drivers with speeding points aren't more dangerous says Safe Speed. Online: Safe Speed press release, 26 April, available at: www.safespeed.org.uk/forum/viewtopic.php?f=14&t=6914&start=20 (accessed 18 August 2010).

Soole, D., Lennon, A. and Watson, B. 2008 Driver perceptions of police speed enforcement: Differences between camera-based and non-camera based methods: Results from a qualitative study, Australasian Road Safety Research, Policing and Education Conference, Adelaide, South Australia, 10–12 November 2008. Online, available at: http://eprints.qut.edu.au/17781/ (accessed 13 July 2010).

Sparks, R. 2000. Perspectives on Risk and Penal Politics. In R. Sparks and T. Hope (eds) *Crime Risk and Insecurity*. London: Routledge, 129–45.

Sparks, R. and Hope, T. (eds) 2000. *Crime Risk and Insecurity*. London: Routledge.

Speedcamerafine.com 2007. Get your speeding ticket cancelled in just five minutes! Online, available at: www.ukspeedcamerafine.com/?gclid=COWAz9-n_6ICFdeX2Aod9jpqeg (accessed 22 July 2010).

Stenson, K. 2001. The New Politics of Crime Control. In K. Stenson and R. Sullivan (eds) *Crime, Risk and Justice*. Devon: Willan, 15–28.

Stradling, S.G. 1997. Violators as 'Crash Magnets'. In G.B. Grayson (ed.) *Behavioural Research in Road Safety VII*. Crowthorne: Transport Research Laboratory, 4–9.

Stradling, S. 2005. Speeding Behavior and Collision Involvement in Scottish Car Drivers. In D.A. Hennessy and D.L. Wiesenthal (eds) *Current Trends in Driver Behavior and Traffic Safety Research*. New York: Nova Science Publishers, 111–21.

Stradling, S. 2006. *Midlands' Drivers' Attitude Study*. West Midlands: West Midlands Casualty Reduction Partnership.

Stradling, S. and Campbell, M. 2002. The Effects of Safety Cameras on Drivers, RoSPA 67th Road Safety Congress, Stratford-upon-Avon, March 2002. Online, available at: www.rospa.com/roadsafety/conferences/congress2002/proceedings/stradling.pdf (accessed 18 March 2004).

Stradling, S., Meadows, M. and Beatty, S. 2004. Characteristics and Crash Involvement of Speeding, Violating and Thrill-Seeking Drivers. In T. Rothengatter and R. Huguenin (eds) *Traffic and Transport Psychology: Theory and Application Vol. 2000*. Oxford: Elsevier, 177–92.

Stradling, S. et al. 2003. *The Speeding Driver: Who, How and Why? (Research Findings No. 170/2003)*. Edinburgh: Scottish Executive Research Findings.

Stradling, S. et al. 2008. *Understanding Inappropriate High Speed: A Quantitative Analysis: Road Safety Research Report 93*. London: DfT.

Sullivan, R. 2001. The Schizophrenic State: Neo-Liberal Criminal Justice. In K. Stenson and R. Sullivan (eds) *Crime, Risk and Justice*. Devon: Willan, 29–48.

The Sun 2006. FATSO GATSO speed cam nets £1m ... at roadworks. 18 April, 1.

Sunday Telegraph 2004. These are greed cameras, not speed cameras. Online 11 January, available at: www.telegraph.co.uk/news/uknews/1451370/These-are-greed-cameras-not-speed-cameras.html (accessed 23 August 2011).

Sunshine, J. and Tyler, T. 2003. The role of procedural justice and legitimacy in shaping public support for policing. *Law and Society Review*, 37, 555–89.

Sykes, G. and Matza, D. 1957/2001. Techniques of Neutralization. In J. Muncie, E. McLaughlin and M. Langan (eds) *Criminological Perspectives*. London: Sage, 206–13.

Taylor, M. 2002. Speed is a key factor in crashes. Online: Transport Retort, available at: www.visordown.com/forum/forummessages.asp?v=1&urn=3&utn=105304&umn (accessed 24 August 2011).

Taylor, M., Lynam, D. and Baruya, A. 2000. *The Effects of Drivers' Speed on the Frequency of Road Accidents: Transport Research Laboratory Report 421*. Crowthorne: TRL.

The Telegraph Motoring 2005. How speed generates heat. 30 April, 4–5.

Thwaite, M. 2008. *The Driver's Survival Handbook*. Sheffield: Streetwise Publications.

The Times 2003. Milking the motorist. Online 27 March, available at: www.timesonline.co.uk.tol/comment/article850385.ece (accessed 28 July 2003).

Times Online 2004. The speed cop's next target. Online 23 May, available at: www.timesonline.co.uk/tol/driving/article428998.ece (accessed 23 May 2010).

TLGR Committee 2002. *Road Traffic Speed Ninth Report of Session 2001–2*. London: HMSO.

Transport Research Laboratory 2002. Speed and accidents – Let's put the record straight! *TRL News*, September. Crowthorne: TRL.

Transport2000 2002. Speed camera challenge goes to the High Court. Online: press release, 28 October, available at: www.acttravelwise.org/news/13 (accessed 24 August 2011).

Transport2000 2003. Focus on road safety and speed. Online, previously available at: www.transport2000.org.uk/campaigns (accessed 20 April 2006).

Tyler, T. 1990. *Why People Obey the Law*. London: Yale University Press.

Tyler, T. and Huo, Y. 2002. *Trust in the Law*. New York: Russell Sage Foundation.

Tyler, T. and Lind, E.A. 1988. *The Social Psychology of Procedural Justice*. London: Plenum Press.

UK Driving Secrets 2010. The site they don't want you to visit! Online, available at: www.uk-driving-secrets.com/beat (accessed 22 July 2010).

Urry, J. 2002. Mobility and proximity. *Sociology*, 36(2), 255–74.

Watson, R. 2002. Ethnomethodology and Textual Analysis. In D. Silverman (ed.) *Qualitative Research: Theory, Method and Practice*. London: Sage, 80–98.

Wells, H. 2004. *Speed Cameras. POSTnote, No. 218 (May)*. London: Parliamentary Office of Science and Technology.

Wells, H. and Wills, D. 2009. Individualism and identity resistance to speed cameras in the UK. *Surveillance and Society*, 6(3), 260–64.

Walton, D. and Bathurst, J. 1998. An exploration of the perceptions of the average driver's speed compared to perceived driver safety and driving skill. *Accident Analysis and Prevention*, 30(6), 821–30.

Walton D. and McKeown P.C. 2001. Drivers' biased perceptions of speed and safety campaign messages. *Accident Analysis and Prevention*, 33(25), 629–40.

Whitlock, F.A. 1971. *Death on the Road*. London: Tavistock.

Wilson, C., Willis, C., Hendrikz, J.K., Le Brocque, R. and Bellamy, N. 2010. Speed cameras for the prevention of road traffic injuries and deaths. *Cochrane Database of Systematic Reviews*, Issue 11, 1–70.

Winnett, M. and Wheeler, A. 2002. *Vehicle-Activated Signs – A Large Scale Evaluation (TRL 548 prepared for Road Safety Division DfT)*. London: Department for Transport.

Woodman, P. 2003. Drivers 'unlikely to shop speed camera vandals'. Online: *PA Consulting News* press release, 20 October, previously available at: www.news.scotsman.community/latest.cfm (accessed 20 October 2003).

YouGov 2003. YouGov survey results: Motorists. Online, available at: http://today.yougov.co.uk/sites/today.yougov.co.uk/files/YG-Archives-lif-dTel-Motorists-031201.pdf (accessed 24 August 2011).

Young, J. 1999. *The Exclusive Society*. London: Sage.

Young, J. 2003. Merton with energy, Katz with structure: The sociology of vindictiveness and the criminology of transgression. *Theoretical Criminology*, 7(3), 389–414.

Index